BIG BIOLOGY

US/IBP Synthesis Series

This volume is a contribution to the International Biological Program. The United States' effort was sponsored by the National Academy of Sciences through the National Committee for the IBP. The lead federal agency in providing support for IBP has been the National Science Foundation.

Views expressed in this volume do not necessarily represent those of the National Academy of Sciences or of the National Science Foundation.

US/IBP SYNTHESIS SERIES | 7

BIG BIOLOGY

The US/IBP

W. Frank Blair

Dowden, Hutchinson & Ross, Inc.
Stroudsburg   Pennsylvania

Library of Congress Cataloging in Publication Data

Blair, W. Frank 1912–
  Big biology.
  (US/IBP synthesis series ; 7)
  Bibliography: p.
  Includes index.
  1. International Biological Programme. 2. Biological research—United States. I.
Title. II. Series: U.S./IBP synthesis series ; 7.
QH9.5.B52   574'.07'2073   77-8512
ISBN 0-87933-305-7

Exclusive distributor: **Halsted Press,**
A Division of John Wiley & Sons, Inc.
ISBN; 0-470-99225-5

Foreword

This book is one of a series of volumes reporting results of research by U.S. scientists participating in the International Biological Program (IBP). As one of the 58 nations taking part in the IBP during the period July 1967 to June 1974, the United States organized a number of large, multidisciplinary studies pertinent to the central IBP theme of "the biological basis of productivity and human welfare."

These multidisciplinary studies (Integrated Research Programs), directed toward an understanding of the structure and function of major ecological or human systems, have been a distinctive feature of the U.S. participation in the IBP. Many of the detailed investigations that represent individual contributions to the overall objectives of each Integrated Research Program have been published in the journal literature. The main purpose of this series of books is to accomplish a synthesis of the many contributions for each principal program and thus answer the larger questions pertinent to the structure and function of the major systems that have been studied.

Publications Committee: US/IBP
Gabriel Lasker
Robert B. Platt
Frederick E. Smith
W. Frank Blair, Chairman

Preface

The more than 12 years of planning, launching, implementation, and wrap-up of the International Biological Program have been an interesting, exciting, and sometimes frustrating period. As one whose fate it has been to be close to the action throughout virtually this entire period, I have written my own account of the IBP as I saw it. This is one man's biased account of the IBP. Others may disagree with some of my generalizations. I have attempted to include the negative as well as the positive, the failures as well as the successes. By recourse to the bales of paper generated by the IBP, both nationally and internationally, I have attempted to make this as factually accurate a history of the IBP as possible.

The various kinds of activities that went on under the IBP logically divide up into a series of chapter headings. First, there were the years of planning a national program of participation in the IBP and the interfacing of this planning with International plans. There was the long-continuing effort, beginning in 1967, to promote participation of Latin American countries in the IBP. There was the tough job of selling IBP as a national program and thus of obtaining Federal financing for it. There was my participation in international activities and in interactions with other international undertakings. The operational phase will receive relatively scant treatment here, as the main purpose of this account is to record the history leading up to the establishment of the IBP as an operational program in the United States. Also in this account, I have attempted my own general assessment of the accomplishments of the IBP both nationally and internationally. Finally, this account is intended as an introduction to the series of approximately 20 synthesis volumes on which the successes or failures of the IBP effort in the United States can be largely judged.

W. Frank Blair
Austin, Texas

CONTENTS

x   Contents

1

The Planning Years: 1962-1967

My first real awareness of the IBP came in mid-March, 1963. I was on the Davis campus of the University of California to give a series of lectures. One day, Ledyard Stebbins invited me to his office in one of the old World War II barracks buildings to tell me about a new international program in biology that was being promoted by the International Union of Biological Sciences (IUBS), one of the unions of the International Council of Scientific Unions (ICSU). Stebbins, who was then Secretary General of IUBS, gave me a reprint of an article he had published in September 1962, in the Plant Science Bulletin concerning the activities of IUBS and particularly concerning its embryonic International Biological Program (Stebbins, 1962a). He had also written a parallel article for the December 1962 issue of the AIBS Bulletin (Stebbins, 1962b). As Stebbins talked about the potential of this as yet vaguely conceptualized international project, I had no forewarning of the depth of my own involvement in it over the next 10 years.

Efforts to initiate an International Biological Program have been generally attributed to the success of the International Geophysical Year (IGY). Stebbins (1962a) credited a meeting of U.S. biologists convened by Drs. Hiden Cox and John Olive of AIBS in 1960 and a subsequent television program in 1961 with stimulating interest in an international biology program on the part of U.S. scientists. C. H. Waddington, who was himself much involved in the early days of planning, credits Sir Rudolph Peters of the United Kingdom with being the first to advance the idea of an IBP in the late 1950's, when he was president of ICSU (Worthington, 1975). Giuseppe Montalenti of the University of Rome, who preceded Waddington as president of IUBS, drafted the first formal plans for IBP, with aid from a preparatory committee appointed by ICSU. This committee produced a three-parted restricted program consisting of : (1) Human Heredity, (2) Plant Genetics and Breeding, and (3) Studies of Natural Plant Communities Which Are Liable to Undergo Modification or Destruction.

The preliminary plans for the IBP were presented to the General Assembly of IUBS in Amsterdam in July, 1961. Following a symposium on possible topics for the IBP, about which Waddington later wrote, "My private opinion was that it would probably be most satisfactory to kill the whole thing before it went any further if this could still be done" (Worthington, 1975). The Assembly passed a resolution calling for careful consideration of an International Biological Programme by a special committee, with suggested research topics being:

"(1) Investigation of biological communities menaced by transformation or destruction.

(2) Changing pressure exerted by the environment on the genetic constitution of mankind and on the dynamics of populations.

(3) Encouragement given to research directed toward a better understanding of and various means of modifying the global equilibrium and balance sheet of organic materials and resources, so as to improve the balance sheet by increasing production and reducing losses."

After approval of this resolution by the ICSU General Assembly in September 1961, a planning committee was organized, headed by Montalenti. This committee adopted the theme, "The Biological Basis of Productivity and Human Welfare," and this theme persisted through the life of the IBP although projects never dreamed of by the planning committee were later to evolve. The committee established six subdivisions, with a separate subcommittee for each. Ledyard Stebbins was a member of the Planning Committee and hence undertook to publicize the IBP planning in the United States.

One of the most active of the subcommittees was chaired by H. Ellenberg, of Institute de Gëobotanique, Zurich. This committee met in Rome in early November 1962, and produced a program paper entitled, "General Survey of Terrestrial Biological Communities dans le cadre du International Biological Programme (IBP)." This proposal was of such nature that I referred to it in later correspondence with C. H. Waddington as an "ill-advised proposal for world-wide quadrat studies." It also had a proposed budget of $25,000,000 (U.S.). The image of the proposed IBP created by this proposal influenced my own negative view of the IBP in the early planning years, and I am sure that it accounted for the negative view expressed by various prominent American ecologists. A quote from correspondence by James Ebert (Worthington, 1975) reflects this view: "at the AIBS meeting last August, most ... of the proposals were for warmed-over ecology of the thirties and forties on a world-wide scale."

In April 1963, I wrote as President of the Ecological Society of America (ESA) to ask Eugene Odum to represent ESA in contacting Ellenberg for a copy of his program and in chairing a committee to examine this program and to make recommendations to the society concerning what should be its stand on this program. William S. Benninghoff of the University of Michigan and R. H. Whitaker, then of Brooklyn College, agreed to serve with him. Odum was a member of the Ellenberg committee, but unfortunately had not attended the Rome meeting.

By mid-1963, various documents were being received from Europe setting forth plans for the proposed IBP. A document describing the proposed IBP was circulated in June 1963 by S. Horstadius of Sweden, who was then President of ICSU. This included a progress report of the several subcommittees, and also emphasized the IBP objectives of productivity and human adaptability. In July, a mimeographed document querying interest in secondary productivity research was circulated by K. Petrusewicz of Warsaw. In July, I also received a form letter from C. H. Waddington and a copy of an article describing the IBP that he had published in the *New Scientist*. The letter read as follows:

"I am sending you a copy of an article I recently wrote about the proposed International Biological Programme, which I am also sending to a number of other senior American biologists who I know to be interested in the application of biology to human welfare. I think that, if this programme is to be worthwhile, it will need the guidance of people who both have some vision of what biology can do for mankind and who can also keep their feet on the ground and prevent the programme going off into mere accumulation of details.

"I found myself somewhat thrown into the middle of this situation when, to my surprise, I was elected President of the International Union of Biological Sciences. I found that they had already in existence a project for an IBP, but it seemed to be extremely vague in its objectives, each member of the Planning Committee suggesting that it should be about his own special field. I tried to look at the situation from an outside point of view. It seemed clear that there was not much case for organised international programmes in the growing points of analytical biology, such as molecular genetics or central nervous system work. The major international biological problem seemed to be "the population explosion"; and in this the control of human reproduction is already the concern of a good number of international bodies. I came to the conclusion that the major contribution which general biologists can make to world progress at the present time is to study the conditions of biological productivity, on which all food production is based, and to force this problem on public attention, trying to make governments and peoples realise that the surroundings in which man lives are a working biological system and not merely something to be exploited as fast as possible.

"To ask for a world-wide survey of biological productivity is, in fact, to face ecologists with a challenge which it is not quite clear that they can meet; those dealing with marine and fresh-water habitats seem to have a fairly well developed science of production, but production by terrestrial communities is much more complex and less well understood. I think it is possible that it will turn out that biologists cannot yet give an account of what living systems in all regions of the world's surface do with the incident energy which falls on them. But even if this pessimistic opinion is correct, at least it will surely be of great value to focus attention on the problem."

I responded to the Waddington communication as follows:

"I am very glad to have the information you recently send concerning the present status of the International Biological Program. The Ecological Society of America is keenly interested in developments in the planning of this program, but we have suffered from poor communications. Until quite recently we had no hard information beyond the ill-advised Ellenberg proposal for world-wide quadrat studies. Speaking personally, I strongly favor the central theme of "biological productivity" as the most promising one for an international effort, and I believe that this is a consensus of leaders among American ecologists. I might state frankly also that a good many people are going to have to be convinced that a sufficiently meritorious project is being undertaken to justify the diversion of manpower and research dollars that would be required for a 5-year effort."

During this time, Jerry Olson of Oak Ridge National Laboratory was one of the main U.S. contacts with developments in Europe. He attended a meeting of the British Ecological Society in March, and at that time talked with Ellenberg. Later, he circulated from Monks Wood Experiment Station, where he was on leave, a progress report on IBP (in September) a paper with J. D. Ovington and H. A. I. Madgwick, defining various terms in production ecology.

In November 1963, an *ad hoc* U.S. Committee on the International Biological Program was appointed by Dr. Frederick Seitz, President of the National Academy of Sciences. Charges to the committee were:

"The function of the Committee for IBP will be to review and evaluate the proposed program in relation to the interests of U.S. scientists, and to make recommendations as to its modification, to identify individuals and groups that might wish to be involved, to arrive at conclusions as to the nature and probable extent of U.S.

participation and to formulate recommendations as to the organizational structure necessary to ensure effective co-ordination of project activities."

Members of the committee were:

Name	Field
1. CAIN, Stanley A. (Chairman) University of Michigan	Conservation Botany
2. BLAIR, W. Frank University of Texas	Animal Ecology
3. CANTLON, John E. Michigan State University	Plant Ecology
4. DAVIS, George K. University of Florida	Biological Chemistry Nutrition
5. DAVIS, Kingsley University of California (Berkeley)	Sociology Demography
6. KETCHUM, B. H. Woods Hole Ocean. Inst.	Marine Biology
7. KRAMER, Paul J. Duke University and AIBS	Plant Physiology
8. LAUGHLIN, William S. University of Wisconsin	Physical Anthropology
9. PARK, Thomas University of Chicago	Zoology Ecology
10. ROBINSON, Sid Indiana University	Physiology

David Frey, Limnologist of Indiana University, was added later.

The report of the international Planning Committee for the IBP had been approved earlier by the Executive Committee of IUBS and in mid-November by the General Assembly of ICSU. This was the chief document to which our U.S. *ad hoc* committee was expected to respond.

The *ad hoc* committee held four meetings between early December 1963 and mid-May 1964. One difficult task was that of ascertaining the amount of interest among pertinent American scientists in becoming involved in the program. One approach was to send a questionaire and a copy of the ICSU proposal to selectedts who had been named by members of the committee. Another was to solicit responses from pertinent scientific societies such as the Ecological Society of America. These procedures, not unexpectedly, produced a mixed bag of responses.

An IBP committee set up by the American Society of Agronomy responded negatively:

> "This committee has reviewed the IBP proposals and finds the viewpoints too restricted and academic to justify ASA participation. If the approach finally adopted by IBP is broadened to permit a realistic emphasis on modern agronomic techniques, ASA might well consider some degree of participation."

I had asked W. S. Benninghoff to poll the Ecological Society of America. By the time of the second meeting of our *ad hoc* committee in early February 1964, he had received 104 replies. From these, he drew the following tentative conclusions:

"1.   In spite of the paucity of definitive information about the IBP available to our members, more than 100 are eager to participate, as that number made the effort to return questionnaires. Not all of those will necessarily find their presently envisioned niches in the program when it is finally formulated, of course; but these and additional members may well find other ways of contributing. The size of response from those wishing to participate also tacitly indicates that a considerable additional number of ecologists are interested in the development and products of the program.

"2.   Specialists in terrestrial ecology are in the majority within the responding group; but there is also considerable representation from freshwater, fisheries, and marine littoral specialists. These facts indicate the subject areas of primary responsibility for the Society, and need for coordination of effort with organizations concerned especially with aquatic biology (e.g., the American Society of Limnology and Oceanography).

"3.   Only about one-fourth of the respondents consider themselves (or the groups of workers they represented) as ecosystem specialists; and more than half of these are attached to three centers (University of Georgia, Oak Ridge, and Hanford).

"4.   The number of specialized centers of existing information and facilities for processing or storing information that were tentatively offered for potential service to the IBP effort make it clear that a careful national inventory is necessary and should be followed by communications with responsible authorities to determine actual availabilities for IBP-associated projects. This would be an appropriate undertaking for the National Committee.

"5.   There may be an opportunity for the Society to assist Latin American ecologists and biological organizations to obtain information until such time as their national directing groups are organized."

I subsequently made my own analysis of the results of the Benninghoff poll and summarized my conclusion for the committee as follows:

"1.   A relatively few centers have expressed interest in IBP — these are principally supported by one government agency, AEC.

"2.   Most of the individuals seemingly will continue to do what they have been doing but under aegis of IBP. If this is so, is IBP worth the time, effort, and money that will have to go into it?

"3.   A considerable increase in personnel and in funding is indicated. From where will this increase come, and will it adversely affect other interests?

"4.   My personal feeling is that the stated objectives of IBP are meritorious. However, I also feel that the proposed plan for the part of IBP I feel competent to judge (terrestrial ecology) cannot attain these objectives and that the contribution will be so minuscule as to place the field of environmental biology in a bad light, particularly if there is a great ballyhoo for IBP.

"My own feeling is that IBP could come the closest to achieving its objectives if an imaginative approach were taken to investigating the functioning of representative ecosystems. I have made this suggestion to Professor Petrusewicz of the subcommittee on secondary production."

Various letters, some solicited but most unsolicited, reached me or the committee. One very critical and pessimistic letter came from Lamont Cole of Cornell University. It read:

"I am writing to let you know my feelings about the International Biological Programme.

"Frankly, I am quite uneasy about this program and the possible involvement of the Ecological Society of America. I fear that a widely publicized effort like this could cause unfavorable reactions to biology in general, and ecology in particular.

"I have heard enough discussion of IBP to be certain that two suspicions are widely held. These are:

a) It is a boondoogle designed to ride the coattails of IGY.

b) It is a scheme to raid the U.S. Treasury, largely for the benefit of foreign scientists.

"I am not concerned with the truth or falsity of these views. The important point is that they will appear to be confirmed if the program is undertaken and is not an outstanding success. In any case, there seems to be nothing in IBP that would require international cooperation. Unlike IGY, for which it was necessary to have simultaneous observations in different regions, the IBP plans do not even attempt to make a case for simultaneous studies.

"I have seen bits and pieces of the European planning of IBP. I suppose we should be gratified that the most important problem they could think of to work on was an ecological one; however, they certainly did not go to Europe's top experts on productivity for that proposal. Then, when it was sent to this country, it got into the hands of the National Academy of Sciences. I suppose it is logical for those unfamiliar with the situation to assume that NAS can speak for all science, but, unfortunately, ecology is a field in which NAS is not highly competent, and in this case apparently didn't know where to turn. I think it is most unfortunate that the problem was not at least handed to the two ecologists (A. E. Emerson and G. E. Hutchinson) who are members of the Academy.

"Now the Ecological Society has been called in, with a timetable already set and a broad outline developed. The ecologists are likely to get the blame if this program is undertaken, underwritten, and unsuccessful.

"My only knowledge of the current status of IBP comes from perusal of the 'Report of the planning committee' dated 15th November, 1963. I have no comment on the section on 'Human Adaptability,' but the remainder of the report just does not impress me as a first-rate ecological proposal. To be specific:

p. 1.   I do not see in what way the program 'will benefit from international collaboration' beyond the fact that we always like to see more people attempting to solve important problems.

p. 3.   They expect 'that the increase in research actually taking place will be modest ...' I agree; the persons working on these problems will continue to do so, and a few grants could attract more into the field.

p. 5.   I don't think 'dry matter' production or 'biomass' change are fundamental parameters. They could at least propose to equip the workers with calorimeters and see where the energy is going.

p. 6.   The microflora and microfauna of the soil are practically unknown. In my opinion, a program to inspire a complete taxonomic investigation of this field would be worth more than the entire IBP proposal. While the free-living mites, nematodes, rotifers, tardigrades, etc. and authentic 'micro-organisms' remain undescribed, we are unable to apply methods of community analysis to the soil.

p. 6.   'Existing meteorological networks' are designed to measure environments that are *not* those of the organisms involved in productivity. Establishment of a genuine microclimatological network would be more valuable than the proposals in this report. (Incidentally, I am amused by the offhand way in which they state that 'evaporation' will be measured 'where necessary'; a standard way of doing this would be a real contribution).

pp. 8-9.   The nitrogen cycle is of undoubted importance. When it is studied in the field, 'nitrogen balance sheets' typically do not balance. The reasons are that we don't know how to measure denitrification, and, as recognized on p. 9, that data on non-symbiotic fixation are scanty, 'due to a lack of suitable techniques hitherto.' I can conceive that the use of N^{15} techniques might help to increase our understanding of both processes—but this would be a basic research problem! If each investigator is going to need a mass spectrograph, it is hard to think of a more improbable program for international collaboration.

p. 9.   I consider the conclusion that 'radiant energy is the ultimate limiting factor for maximum productivity' to be highly questionable.

I am sorry that, in the above, I have dwelt on trivia, but I am commenting on the only information available to me about this program. It impresses me as contrived, and more likely to damage than to help the image of ecology if the impression gets around that this is the best research plan ecologists can produce. A proposal for an international program should be authoritative and persuasive, but this reads like a run-of-the-mill request for a grant.

"If the Programme goes ahead I would not try to discourage ecologists from participating, but I would hate to see the Ecological Society endorse it officially as though it were a plan of our own. When and if it becomes evident that an International Ecological Program is really needed, we should join with our sister societies in Great Britain and Japan to identify the problems most in need of attack, and to devise a really up-to-date program for dealing with them."

Another negative letter came from Nelson Hairston of the University of Michigan, who has remained critical of the IBP throughout its entire history. His letter read:

"A group of biologists at the University of Michigan met recently at the request of Professors Stanley A. Cain and James V. Neel, to be informed of the plans and organization of the proposed International Biological Program (IBP). From information obtained at the meeting it appears that you will be in a position of influence with respect to the proposal, and I am therefore writing a personal letter to you concerning some of the reservations that I have concerning the IBP.

"It seems to me that we should ask ourselves the following questions:

1. Do the aims of the program, as stated in the Preamble of the Report of the Planning Committee, make the proposal worthwhile?
2. Would the program result in an important research that would not otherwise be done?
3. Are there any benefits that will follow from the program, other than those implied by positive answers to questions 1 and 2?
4. Would support of the program by this country mean that new funds will become available for the type of research that is envisioned?
5. Do the possible benefits outweigh the cost in organizational effort and funds?
6. Is the composition of the various IBP subcommittees such that you would wish your own research program 'coordinated' by the relevant committee?
7. Of what would 'coordination' consist?

"I think that you will agree that a negative answer to any of Questions 1, 2, 4, 5, or 6 would cause one to have serious reservations about supporting the IBP, and that negative answers to more than one of these questions would cause one to oppose the program. It seems to me that several of the questions cannot honestly be given a 'yes' answer, although they might not be answered with a definite 'no'. These are questions 2, 4, 5, and 6, and brief reasons for my opinion follow.

"2. In my opinion, the amount and quality of research done at present is limited by the number of able people and the amount of time that they can put into research effort. If this is true, no amount of international consultation can increase the research output within the time envisioned. Indeed, there might be an inverse relationship between the answers to questions 2 and 6. If the more able people, such as Odum, spend enough of their time or coordinate the effort, their own output would of necessity decline.

"4. We have been assured that each country would pay for its own part in the program, but this statement ignores (among other things) the fact that the specialized agencies of the United Nations receive more than 1/3 of their operating funds from the United States. WHO is one such agency about which I have first-hand knowledge. The U.S. pays about 32% of the regular budget, and in addition contributes considerable extra funds through NIH grants and special donations to specific programs, such as the Malaria Eradication Campaign. These United Nations agencies are very unlikely to support any American participation in IBP.

"More serious is the question of possible diversion of funds from our regular granting agencies, such as the National Science Foundation and the National Institutes of Health. It is clear that if no special appropriation is forthcoming, it will become more difficult to obtain support for research not oriented towards IBP. The recent history of the Congress is particularly non-reassuring in this regard.

"5. The meeting referred to at the beginning of this letter was attended by around 30 people, and lasted for 1½ hours, thus consuming more than one week's time for one person. If one considers the number of such meetings that are likely to be held, the total time consumed impresses one deeply, and this time is only to pass information to interested people. We should also consider the time already spent and to be spent in the rapidly multiplying committees, subcommittees, and *ad hoc* committees in various societies, universities, and national and international groups. The total loss to research time is appalling if able people are involved, and if the committee membership are given to people who are not first rate, the results of their efforts will be equally so.

"6. To speak frankly, the membership of some of the committees is not at all reassuring, particularly with regard to American representation, or lack of it. It is inevitable that the major effort would come from this country, and it is only appropriate that American biologists should be represented in a much better fashion.

"I believe that my record of work with WHO (over 4 years out of the last 10) is sufficient evidence that my attitude towards IBP is not based on isolationism or chauvinism, but comes from a concern for the proper use of personnel and funds in the advancement of biology in general and ecology in particular.

"It is my conclusion that the proposed program is an unfortunate example of 'me too', and will not result in any worthwhile gains."

R. S. Miller, then at the University of Saskatchewan, was the only North American to participate in a Paris meeting of Ellenberg's subcommittee in mid-December, 1963. At my request Miller provided me with the following resume of the meeting:

"1. Meeting was called to consider ways and means of undertaking an international, cooperative study of secondary productivity with selected groups of animals in selected habitats.

2. Soil ecologists were the only group prepared to agree on study areas, animals to be studied or methods of study; although they felt that an introductory and comparative study of extraction and analytical techniques would be required before usable data could be obtained.

3. Vertebrate and invertebrate ecologists (other than soil zoologists) described their previous and current research, but there was no indication that they were willing or able to adjust their methods, research aims, or emphases to fit an international program.

4. There was no agreement on methods of obtaining or analyzing population estimates (esp. vertebrates); in fact, most individuals seemed to stress his own method as the best, even though there was obvious disagreement about the accuracy of different methods.

5. Very few of those present understood what was meant by 'secondary productivity' and, conversely, those who did (e.g., Petrusewicz), failed to appreciate the errors and variables involved in population estimates.

6.   There was agreement among soil ecologists that grassland habitats would provide data for the most reasonable comparisons between countries or areas, but others reached no agreement or understanding on this point.

7.   None of the representatives had any prospect of funds for his research, and as near as I could ascertain, no national government or agency had committed itself to research support—I was left with the impression (although it was never quite stated) that if a proposal could be drafted, the U.S. would provide funds.

8.   For the most part, the impression I had was that IBP might be a source of research funds (e.g., U.S. source), but not that IBP was a necessarily worthy project in itself. The members at this meeting did not approach IBP from the point of an ecological problem that needed to be solved, or was worth studying, but rather looked upon IBP as a bandwagon they might jump on, *if* it would support their personal research intents."

Some of the unsolicited comments coming to our committee are represented by the following excerpts:

"Perhaps because my activities have been too largely restricted to a sub-basic branch of biology in which problems of international cooperation are minimal, I find myself sympathizing with those few members of the Division of Biology and Agriculture who question the desirability of an International Biological Program. It seems to me that the aims of an IBP should be achieved under the auspices of one or more of the organizations now existent. A new program would only drain specialists from other programs that are already undermanned. In my opinion all too many of our biologists are even now devoting a disproportionate amount of time to conferences at the expense of individual research. If more of these outstanding intellects were freed of conference commitments, is it not possible that our quote of Darwins and Einsteins might be increased. In other words, I do not believe that the obvious propaganda value of an IBP is commensurate with the manpower cost that would be involved.

"I wholeheartedly concur in the sentiments you expressed in the first paragraph of that letter. The International Biological Program is, in my opinion, unacceptably reminiscent of "'me-too'ism". It took care of promoting gentry in biology to get on the international gravy train. Secondly, and the point which you made, the functions proposed fall into areas already in the cognizance of FAO, WHO, etc. Thirdly, the IBP would represent an unjustified drain on available manpower. Fourthly, the existing agencies having cognizance are already manned and operating. Needless duplication of manpower and effort would be created by IBP. Fifthly, a fraction of the money envisioned for IBP if put into the existing agencies would perform the function of IBP ..."

Even Tom Park, a member of our *ad hoc* committee, wrote a cautionary memorandum stating his personal reservations about the IBP as follows:

"I wish to record here some of my reservations regarding the IBP. I do not do this to be destructive or petty. Rather, it is my hope that the American(s) sent to the Paris meeting this summer will be critical and questioning rather than complimentary and applauding. My points are these:

1. If there had not been an IGY there would not now be projected an IBP.
2. The IGY had meaning. Relatively simple measurements could be defined and taken. These data, in turn, contributed *conceptually* to geophysics.
3. The IBP does not enjoy this meaning:
   (a) No such simple measurements can be taken.
   (b) There is little, if any, conceptual framework onto which the measurements that are taken can be apportioned.
4. The IBP is not a *biological* program in any comprehensive sense. It contains a part of ecology (and not the best part in my biased view); a limited part of physiology; a very small part of genetics and nutrition; and a rather non-rigorous treatment of conservation. A large segment of biology is totally omitted. At the very least, and for several reasons, the name of the Program should be changed.
5. Personally, I believe that Biology, and even the study areas now included in the plan, would be better off in the years to come if the IBP were not to be held: If science evolved in its own way without the seductive pressures of funding (largely U.S.A.?), organization charts, preconceived concepts, and 'glamour.'
6. I fear that much of the work that will be done may prove to be slipshod and uncritical. There will be much breadth but little depth, with the notable exceptions of those investigators who would do their work in any event."

"Despite my reluctance about the Program it seems evident that it will be initiated. This being so, the United States is more or less compelled to joint the enterprise. The most constructive things our Committee can do are (i) to provide the Paris delegate(s) with a full treatment of our views and those of our consultants; (ii) to suggest a new name; (iii) to suggest a revised orientation if we can come up with a better one; and (iv) to nominate to the Academy the best possible person(s) to represent both biology and this country at the Paris meeting.

"I will give further thought to point (iii) above."

"Not all of the input to the *ad hoc* committee was unfavorable."

Dennis Chitty of the University of British Columbia responded to my request with a statement endorsing the IBP in principle but pointing up the need for better focus of the proposed research. Bostwick Ketchum of our committee sent questionaires to 37 individuals, most of whom were doing research on the productivity of marine communities. About 75% of those who responded were favorable and indicated that they would participate. Excerpts from the response to the question of "Should U.S. scientists participate?" give an idea of the generally favorable reception by marine biologists:

"I find the Program as a whole very interesting, important, and timely. We would certainly want to participate ... It is a most exciting program. Thanks for letting me comment."

"It seems to me that the program as a whole is well conceived and of the greatest importance."

"There is, however, another important aspect — the "advertising" of the usefulness and scientific respectability of taxonomic and field ecological studies."

"If IBP can bring home to governments the need for research into the basic life-processes of the sea then I am all for it ..."

'The plan looks very good. Perhaps it is too ambitious but it should be worthwhile even if only part of it is accomplished."

"Part C of the report on an IBP seems to be. an outline of what we are doing out here as we can. Thus, how can I argue?

"My first impression after reading the committee's report was 'Fine, this is what most marine biologists and biological oceanographers are now in the process of doing.' ... It is hard for me to believe that a biologist could have no interest in the IBP Committee report."

"The program for IBP as such is very wide and the wording general enough so that everyone can find something in the program to which he could contribute."

"IBP is the sort of thing I would like to see around provided it didn't get into my hair and was useful to me when I wanted it. But then, that is what we all want of our international organizations and it is not good enough."

"... I fear that the plan may be too broad. I wonder whether re-naming projects and the best of coordination can significantly improve the overall effort."

"If IBP is to have any integrity and meaning, it should be definitive and strictly limited in scope and objectives. As proposed, it differs in no way from the entire field of biological oceanography (= marine biology) and will attract neither new scientists nor new funds. I cannot see what purpose can be served by putting a new label on old goods, particularly when the latter are perfectly saleable under their present designation."

"I wonder whether there are enough biological oceanographers around for manning any more programs than are at hand now?"

"... I view all sorts of worthy schemes with a jaundiced eye ... we just do not have the people to do this job — and at some institutions we have people who should *not* be doing the job."

"... none of us feel that we would benefit from or would desire to engage in any new research programs developed purely for cooperative purposes."

"... there are more and more organizations and more and more university meetings to an extent that cuts severely into the scientific working time of a great many people who might be more productive if they did not get enmeshed in these affairs."

"I do not notice that people in oceanography and biological science lack for the ability to travel to foreign countries and to cooperate with their co-workers."

"... the IBP obviously is being conceived as the biologist's IGY, the types of problems are very different."

"I fail to see why the proposed study of biological productivity requires simultaneous observations."

"The problems of marine productivity are going to require generations of effort. To suppose that any significant progress will be made during the course of a five year wing-ding is optimistic, and I can visualize some disadvantages such as (a) creation of over-optimistic hopes in the mind of the public, who might think we are going to double the world's food supply or something; and (b) accumulation of alot of junk data that will be scrapped when the next 'standard' method comes along."

"... these international organizations are sterile and useless if they do not have direct contact with and command the respect of the working scientist ..."

"Wouldn't it be a better idea for the drafters of these programs to persuade any remaining scientists connected with IUBS, ICSU, and SCOR, etc. to get back to the bench and cease from stressing the obvious so much. There is nothing new in this melange."

"My attitude at the moment is 'interested but wary'—which seems to parallel that of your committee!"

My own reservations about IBP as it had been planned up to early 1964 were reinforced by the general tenor of the responses our *ad hoc* committee was getting from the American scientific community. In February, I responded to the document I had received earlier from Petrusewicz proposing secondary production studies in the IBP. I quote my letter in part:

"My own feelings are that relatively haphazard studies of secondary production at a great many stations even if restricted to micromammalia and possibly herbivorous insects are going to profit little from international effort and are going to result in little in the way of worthwhile information. This brings us back to your Proposal I, and I can see this as having the greatest promise. In other words, it seems to me that one of the greatest possible contributions IBP could make would be to further the development of a few centers directed toward a real understanding of how an ecosystem functions. This would mean focusing on high development of the selected centers, with investigations being undertaken on production at all trophic levels. Perhaps a latitudinal transect involving 4 to 5 centers in the landmasses of the Old World and a similar number in the landmasses of the New World would be feasible. Concentration on this proposed plan would not rule out other investigations of secondary production, but I do feel that possible contributions from these other efforts would be relatively minor by comparison with the possible contributions from the two chains of major stations devoted to an understanding of the dynamics of representative, total ecosystems."

Petrusewicz responded to this with a partial acceptance of my suggestion. His letter is quoted in part:

"As to your suggestions, I personally think that in the future IBP should deviate into two lines: (1) world wide investigations on chosen elements of secondary production. It seems to me that it would be very valuable to investigate by comparable methods the extent of secondary production, for instance, of small rodents or other ecologically separate units. Such investigations should naturally include various geographical zones with different species composition. The investigations of certain elements of secondary production will be reasonable providing that (a) a considerable number of habitats is studied and (b) the investigations concern relatively distinct, ecological units of the similar niche.

"(2) investigations of the type you are proposing, my Proposal I, that is, possibly whole ecosystems should be investigated at several, at least, centres. This matter will be discussed after obtaining preliminary results of the preparatory investigations. At present, some difficulties arose at (1) establishing an economic and reliable method of recording numbers of animals in field and (2) at converting numbers, biomass, into production, especially defining turnover."

As time for drafting the report of the U.S. *ad hoc* committee approached, Stanley Cain, Chairman, asked for reports from individual members. My response was so negative that I presumed, wrongly, that this would end my association with the IBP. Main points of my letter to Cain were:

"Does the committee feel that the results to date suggest formal U.S. participation in IBP? The answer here is NO both in respect to my interpretation of the feelings of the large majority of the committee and in respect to my own feelings.

"This does not mean that I am opposed to international cooperation in research in biology. I think such cooperation is highly desirable whenever and wherever good reason for it exists. I practice it myself in my own research, and at present I work in close cooperation with various foreign workers, particularly in Latin America. However, the IUBS proposal simply has not made a case for the proposed international program. It is clearly put forward that people will continue to do what they have been doing but under the umbrella of IBP. This is a contrived concept of international cooperation. Despite denials, it is clear that IBP represents an effort to ride the coattails of IGY. Despite statements to the contrary, there is evidence that IBP is regarded by many in Europe as a contrivance to promote support of their research by American dollars. The one person from Great Britain with whom I have talked stated this quite bluntly.

"Results of efforts to ascertain attitudes of American biologists. Statistical: The only extensive survey is that undertaken by the Ecological Society (refer to my report of March 10). I do not interpret these results as indicative of strong support for IBP among American ecologists. Lots of people wouldn't object to getting more support for continuation of work they already are doing.

"Specific activities suggested. This runs the gamut of ecological research, since each person expects to continue doing what he is already doing. No enumeration seems necessary.

"Personal recommendation. My personal position is that the U.S. delegation would be instructed to state quite bluntly the American objections to the program as it has been approved by IUBS and ICSU.

"A counter proposal is probably in order, reflecting the general feeling among American marine, freshwater, and terrestrial ecologists that there might be justification for high development of a few centers directed toward functional studies of ecosystems and serving as centers for testing and development of methods of population and community ecology which we simply do not have adequately developed at this stage of the game. International cooperation would involve exchange of personnel, training of students, and intercalibration of methods. Choice of stations for development would involve sticky problems of biopolitics and should be done with deliberation and care. We should not rush into IBP if we go in at all. Poland already has a program on productivity under way (refer to previously circulated material from Petrusewicz) in spite of their lip services to cooperation and coordination."

The *ad hoc* committee completed its report by mid-1964. The report was favorable to U.S. participation in the IBP. The recommendations to the National Academy were as follows:

"There appears to be sufficient American interest in the International Biological Program which has been proposed by the International Council of Scientific Unions and approved by the International Union of Biological Sciences (draft of 15 November 1963) to warrant participation by scientists of the United States. It is recommended, therefore, that the United States participate and that the National Academy of Sciences-National Research Council take responsibility for the following steps or organize, coordinate and, as may be necessary, administer the American effort by:

Sending an American delegation of scientists to the Paris meeting of the International Council for the purpose of negotiating changes in the proposed program so as to make it one to which Americans can subscribe and give enthusiastic participation;

Forming a committee of the NAS-NRC to organize, facilitate and coordinate the American effort nationally and in cooperation with the international program;

Seeking postponement of the official developed part of the program until there has been sufficient time for thinking through the objectives and planning the operations, even though this should require a few years of preparation; and,

In view of the scope and heterogeneity of the probable program, by working through a series of subcommittees constituted for clearly defined segments of the program."

A 10-man delegation representing NAS/NRC and co-chaired by Ted Byerly and Stanley Cain participated in the ICSU/IBP meeting in Paris on July 22-25, 1964. Nine other U.S. scientists also participated as representatives of various organizations. Eugene Odum went as a representative of the Ecological Society of America.

The NAS/NRC delegation, on its return, submitted a report that included unanimous agreement on three recommendations. These were:

"1. That the United States, through the National Academy of Sciences, participate in the International Biological Program;
2. That the Academy form a United States National Committee on the International Biological Program; and,
3. That the United States subscribe, through the International Council of Scientific Unions, to the Special Committee for the International Biological Program under the dues category No. 6, which is $10,000 per annum, and assume an appropriate share of additional international costs of the IBP."

These recommendations were accepted, and the United States was committed to participate in the IBP.

With submission of its report, the *ad hoc* committee ceased to exist. As the time was approaching for the naming of the U.S. National Committee for the IBP, there was concern among ecologists about adequate representation of ecology on the committee. In August 1964, the Council of ESA directed Paul Pearson, Secretary of the society, to contact NAS/NRC and make known the interest of the society in making nominations for membership on the committee. Contact was made with Ted Byerly, Chairman, Division of Biology and Agriculture, NAS/NRC, and on his invitation, Eugene Odum, President of ESA, submitted a list after input from Paul Pearson and myself. Contrary to some past experiences with NAS/NRC in ecological matters, the National Committee and subcommittees for the IBP did get a strong representation of ecologists.

The chairman of the U.S. National Committee was made known in January 1965. I and other ecologists were shocked and angry that not only was he not an ecologist, he was not even a biologist. Roger Revelle of Harvard University qualified as one with broad international experience in science, but the ecologists felt that it represented an arrogant disregard for biology on the part of NAS/NRC to place a physical scientist at the head of the U.S. effort in the International Biological Program.

I wrote on January 11 to President Frederick Seitz of NAS/NRC saying:

"I was shocked to hear this morning a rumor that the Chairman of the National Committee for the International Biological Programme was going to be a non-biologist. Without any prejudice to the person named, I feel that it would be most unfortunate to select a person other than one with a broad ecological view to head this committee. My experience with the planning of IBP to date clearly impresses me with the certainty that the program is largely an ecological one. It is very difficult to visualize adequate guidance of the committee by one, however able in other respects, who is not only not an ecologist, but not even a biologist.

"This letter is written simply to express my concern and to assure you that the same concern is going to be felt by a large number of American ecologists."

Seitz responded immediately, saying in part:

"Thank you for your letter of January 11. As you may know, I have been acting under the advice of several groups in connection with the development of the National Committee for the International Biological Programme. I discussed the issue which you raised in your letter with the advisory group before making the appointment, and was assured that the advantages of moving in the direction we did greatly outweighed the disadvantages. There was a strong feeling that if, for a period of a year or a year and a half, the program were placed in the hands of a scientist who is widely conversant with the international scene and a diversity of international problems affecting science, and if he was backed by a strong team of biologists, we would probably move ahead with our part of the International Biological Programme both rapidly and in the right direction."

In late January, the Study Committee of the Ecological Society of America, meeting in Washington, discussed the appointment and directed

ESA president Eugene Odum to write to President Seitz to call to his attention the ESA's interest in the U.S. Committee for IBP and to urge the due representation of research biologists, and especially ecologists on the committee and in the position of vice-chairman.

Ted Byerly and Stanley Cain were named co-vice-chairmen of the 15 member U.S. National Committee for IBP. Seven subcommittees were established conforming to the seven sections (themes) recognized by SCIBP. The first meeting of the committee was held in March and the second in May 1965.

At the second meeting, two additional subcommittees were approved, bringing the number of U.S. subcommittees to nine. One of these was on "Systematics and Biogeography." The second was on "Environmental Physiology." These two committees were added as a result of pressures by prominent leaders of these areas of research, who believed that there was a place for their disciplines in the IBP.

Report One of the USNC/IBP was published in August 1965, under the title, "Preliminary Framework of the U.S. Program of the IBP." This contained preliminary drafts of plans developed by the seven original subcommittees.

When the Environmental Physiology Subcommittee was set up subsequent to the May meeting of USNC/IBP, I was asked to be a member and thus found myself back in the midst of IBP planning and with a more optimistic view of its possibilities for success. In spite of its name, this subcommittee was made up of an interesting mix of evolutionists and biopoliticians in addition to legitimate physiologists. Ted Bullock was chairman, and Ladd Prosser was vice-chairman. Others included Herbert Baker, Theodosius Dobzhansky, Wallace Fenn, Sidney Galler, Colin Pittendrigh, Knut Schmidt-Nielsen, and Carroll Williams.

The last half of 1965 and most of 1966 were a period of usually frustrating planning by the USNC/IBP and its subcommittees without knowing whether there would ever be financial support for the program. Late in 1965, Donald Hornig, Director of the Office of Science and Technology, formally expressed Federal interest in the IBP in a letter to Leland Haworth, Director of the National Science Foundation. The Foundation was asked to be the lead Federal agency for the IBP. An Interagency Coordinating Committee was set up with representatives from pertinent Federal agencies and with Harve Carlson of NSF as its chairman. Carlson's strength within NSF and in "arm-twisting" representatives of other Federal agencies to support the USNC/IBP and its subcommittees with funding were to be a major factor in eventual success of the US/IBP.

There was a great deal of floundering during this period. The meetings of USNC/IBP were large gatherings that were too large to work efficiently. Since the attendance varied, it was typical that the same ground was replowed in meeting after meeting. In retrospect, we were doing at least two things that were hindering the formulation of a viable and attractive U.S. program. For one thing, we were hewing too closely to the framework of the various subcommittees, especially the seven recognized by SCIBP, rather than paying attention to biological problems that needed to be addressed. Secondly, we were prone to try to provide something in the IBP for everyone who wanted to be identified with it.

One of the really wasted efforts during this period was that of identifying as a part of US/IBP individual research projects if they fell within the purview of any one of the nine subcommittees. The minutes of the January 1966 meeting of the USNC/IBP carry the following statement:

> "The USNC/IBP requests Federal agencies to provide inventories of relevant current and planned research,
>
> "The USNC/IBP will welcome proposals for inclusion of specific projects as part of IBP,
>
> "The USNC/IBP shall recognize as part of IBP all projects that result directly from actions by its subcommittees and any other project listed as relevant to its program in categories considered appropriate by the subcommittees."

Part One of Report Three of the USNC/IBP consisted mainly of abstracts of 104 individual research projects that had been identified with one or the other of the nine subcommittees and hence considered a part of US/IBP. By the time part two of this report was published in December 1968, the number had increased to 266. This whole exercise created an image of the IBP that was hard to dispel as we later tried to gain support by both government and scientists.

The general tenor of this period is reflected in a report to the annual meeting of the Ecological Society of America by Ray Fosberg, who was the society's representative on the National Research Council. His reference to IBP read:

> "The most active concern of the Division of Biology and Agriculture in the ecological field has been with the IBP. The U.S. National Committee on IBP and its subcommittees have held a number of meetings. Little seems to have been done, though a statement of guidelines for submission of projects for recognition and approval by the U.S. National Committee for IBP was issued. There still seems to be no provision in sight for financing any US/IBP activities other than committee meetings. IBP activities will, according to the Committee guidelines, have to compete with all other biological proposals for the normally available research money. This seems to indicate that unless the policy changes there will be little additional biological effort over what would normally be carried on."

In spite of everything, though, the concept of whole ecosystem research was slowly gaining acceptance by the U.S. planners. In July 1966, the subcommittee for Productivity of Terrestrial communities presented a revised U.S. program and redefined its objectives as follows:

> "The general objective of the program is conceived to be the study of landscapes as ecosystems. Emphasis is placed not only on the primary productivity of terrestrial vegetation and its meaning for man, but on trophic structure, energy flow pathways (food chains), limiting factors, biogeochemical cycling, species diversity, and other attributes which interact to regulate the structure and function of communities."

By the time that Part One of Report Three was published in June 1967, the Terrestrial and the Freshwater Productivity subcommittees had joined forces to develop an integrated research program on "Analysis of Ecosystems." The turning point that led to this most ambitious of U.S. efforts is generally credited to a large meeting of the USNC/IBP and its subcommittees that was convened at Williamstown, Massachusetts, in October 1966. Fred Smith, then at the University of Michigan, was strongly influential in leading the U.S. efforts in this direction, and he became the Director, Analysis of Ecosystems Integrated Research Program (IRP). I missed this important meeting as I was in Argentina for the I Jornadas Argentinas de Zoologia.

The stated objectives of the Analysis of Ecosystems IRP were:

> "To understand how ecological systems operate with respect to both short-term and long-term processes (systems disturbed to varying degrees by man are included).
> "To analyse interrelationships between land and water systems, so that broad regions may be considered as wholes.
> "To estimate existing and potential plant and animal production in major climatic regions of this country, particularly in relation to human welfare, as our contribution to the worldwide goals of the IBP.
> "To add to the scientific basis of resource management, so that optimization for multiple use and for long-term use can be improved.
> "To establish a scientific base for programs to maintain or improve environmental quality."

Six biome studies—grasslands, deserts, deciduous forests, coniferous forests, tundra, and tropical forests—were proposed for the Analysis of Ecosystems Program. A two-year grant was obtained from NSF for the planning and development of these programs. It was decided to concentrate first on the development of one biome, and the grasslands were chosen as one that showed the most promise of early success. In October 1967, the Pawnee Site in northeastern Colorado, was chosen as the place for intensive study. Colorado State University at Fort Collins became biome headquarters and George Van Dyne became the first biome director in the US/IBP. The Analysis of Ecosystems approach was much more appealing to American ecologists and other environmental scientists than had been anticipated by earlier critics of the IBP.

In the first annual report of the Analysis of Ecosystems Program, Fred Smith wrote:

"The total activity in this program and the number of interested participants have become overwhelming. The decision to have six biome directors derives from the size of the program, which has grown too large for the original staff to handle. It is now evident that this program is becoming a nationwide organization for the development and integration of research on ecosystems. The readiness of ecologists to join in such a venture is much greater than expected.

"The causes of this flood of activity can only be surmised. One idea presented frequently is that everyone is aware that ecosystem problems cannot be solved by individuals. Another is that our scale of study must be expanded to meet the impending crisis between man and the biosphere. Certainly the national concern for environmental quality has influenced the attitudes of many ecologists. Finally, many are aware that systems analysis is much more than a catchword, and offers great hope that complex, collaborative studies will be fruitful.

"Whatever the causes, a revolution among ecologists is underway, and the IBP is in the middle of it."

He also foresaw the need for additional funding if this program was going to succeed.

Other integrated research programs were evolving in 1967. At its late October 1967 meeting, the USNC/IBP accepted nine major projects for inclusion in the US/IBP subject to justification of individual budgets to granting agencies. These were:

Grasslands Biome of the Analysis of Ecosystems IRP.
Study of Eskimo Populations (later split into separate IRP's on Eskimos and Aleuts).
Ecology of Migrant Populations (later to cause many headaches for the USNC/IBP).

Phenology (which failed to develop, partly because its director was in-
capacitated by a heart attack).

Hawaii Terrestrial Biology (later changed to the Island Ecosystem Stability
and Evolution Subprogram of the Origin and Structure of Ecosystems
IRP).

Aerobiology

Convergent and Divergent Evolution (later restructured and renamed the
Origin and Structure of Ecosystems IRP).

Analytical Experimental Biogeography of the Sea (later dropped from the
US/IBP).

Adaptation of Peoples at High Altitudes (later to produce the first synthesis volume in the US/IBP series).

The official start up of Phase II of IBP, the research phase, was
July 1, 1967, for what was planned to be a five-year program. In the
United States, we were still planning. No new research was actually under
way, nor did we know when funds could be found to implement our plans.

The British journal *Nature* took cognizance of the U.S. planning in its
December 2, 1967 issue under the headline "Big Biology."

"Consideration of the philosophical bases of the International Biological Programme (IBP) has so preoccupied scientific
leadership in the United States that observers elsewhere have
wondered when American biologists might actually get down to
specific projects, and what these might be. A stream of stately
essays has been the main output so far; yet Phase 2, or the
operational part of the IBP, was supposed to start last July. At
that time, however, the National Committee was in the midst of
Congressional hearings on the IBP and its funding.

"It now seems that, having scrutinized the underlying
concepts of IBP more thoroughly than any other community, the
United States may now undertake work that is proportionately
more significant. That is clear from the most recent report of the
National Committee, Report No. 3, Part 1, now published. Of
the greatest interest are the major integrated research projects
specially developed within IBP guidelines by the National Committee and directly sponsored by it. The radical sweep of some of
these programmes makes most of the efforts of other countries
look very small beer. None is expected to cost less than $2 million
and several will cost a great deal more."

The Grassland Biome received its first funding for research on June 1,
1968, just 11 months after the official beginning of the research phase of
IBP. The problem now was to organize and shake down the other U.S.
projects and to obtain funding for them.

2

Efforts in Latin America

Efforts to involve Latin American scientists in the IBP and to develop cooperative IBP projects with them had modest beginnings in November 1966 on the occasion of the Primeras Jornadas Argentinas de Zoología. This was held at Horco Mollo, a self-contained convention center in the edge of the submontane cloud forest above San Miguel de Tucumán. The Jornadas had been convened in celebration of the 150th anniversary of Argentine independence. I was an invited speaker principally as a result of my previous collaboration with José Cei of Universidad Nacional de Cuyo in research on toad evolution.

There were informal discussions of the developing IBP with several of the participating Argentine and Brazilian zoologists. The result of these discussions was an informal meeting of those members present from the Asociación Latinoamericana de Ictiólogos y Herpetólogos (ALAIH) and the drafting of a resolution to convoke a symposium in São Paulo, Brazil, in November 1967, with a theme of "Coordination of Studies Relative to the Evolution of the Amphibians and Reptiles of the New World." The intention was that this would be a part of the IBP. The resolution was signed by Dr. Marcos A. Freiberg, president of ALAIH, Dr. Bertha Lutz, premier Brazilian amphibianologist, Drs. Avelino Barrio and José Cei, the two top Argentine herpetologists, R. F. Laurent of Instituto Miguel Lillo, Profesora E. Acosta of Instituto Butuntán in São Paulo, Mr. C. V. Bottari, a student of Cei, and myself.

On returning to the United States, I reported the results of my Argentine visit to the USNC/IBP and requested their approval and financial support for a trip in early 1967 through major Latin American countries to encourage their participation in the IBP. In the committee meeting, this proposal was vigorously opposed by Lee Talbot and Philip Humphrey, both of whom were at that time on the staff of the Smithsonian Institution. The basic reason for their opposition has never been quite clear. At any rate, the pressure from the Smithsonian was so great that I had to go with Mike DeCarlo, NAS/NRC staffer for the IBP Committee to get clearance for the trip from Stanley Cain, then Assistant Secretary of the Department of Agriculture and Vice-Chairman of the USNC/IBP. Stan agreed that the trip should be made, and I proceeded with plans to leave on January 10, 1967.

My plans were to circle South America, visiting 16 cities in seven countries in a period of 25 days. Amazingly, I was able to stick almost precisely with my preplanned schedule, although I travelled on many different airlines, both large and small.

The first stop was in Caracas, where arrangements for my coming had been made by Osvaldo Reig, an acquaintance of several years standing. Reig was one of the many excellent scientists who left Argentina at the time of the overthrow of the government in 1966. The National University in Caracas was occupied by troops in response to an alleged student conspiracy. However, the key group with which I wanted to make contact was located at the Instituto de Biología Tropical, which is some distance from the main campus and was undisturbed by the military. I spent one day in Caracas and met for an extended discussion with about 20 botanists and zoologists at the Instituto. Initially, there was suspicion about the motives for my visit to South America. My motives were questioned principally by Franz Weibezahn of the Departamento de Hidrobiología, but I think I convinced him and the others that the U.S. was not out to exploit them in any way. The reasons for the suspicions became apparent as members of the group began airing their grievances against visiting North America biologists. Two cases were cited, both involving Smithsonian people. These were only hearsay accounts, but they revealed a resentment of U.S. activities that would have to be overcome if effective cooperation with the Latin American scientists were to be established. To the credit of the Smithsonian, the director of the Natural History Museum later took the initiative in promulgating a code of ethics for foreign field work that should help to obviate such resentments in the future.

With the negatives out of the way, I had a very profitable discussion with the Venezuelan scientists. One of the principal concerns expressed was for better communication, especially among the various Latin American countries. For example, Weibezahn, director of the small Instituto de Limnología at Caracas was unaware of the large Instituto Nacional de Limnología at Santa Fe, Argentina. The idea for an IBP Interamerican News, which we were later able to establish and publish through most of the duration of the IBP, was first broached here.

Plans were made for organizing a Venezuelan National Committee for the IBP under Asociación Venezolana para el Avance de la Ciencia (ASOVAC) as there was no Venezuelan National Research Council. This was soon done with Weibezahn as chairman, but, as with most of the other Latin American countries, their active participation in IBP was minimal.

From Caracas, I flew to Belém at the mouth of the Amazon to contact Dalcy Albuquerque, who was then director of Museu Paraense Emilio Goeldi. Albuquerque received me very cordially and spent about an hour talking with me, but, unlike most people I met in South America, he did not go beyond this formal meeting in making me feel welcome. There were undercurrents that pointed up the already known fact that Amazonia is a very sensitive subject in the dealings of Brazilians with foreigners. Soon after I arrived, Albuquerque made a cryptic statement to the effect that, "you can see that I don't eat Americans."

Albuquerque expressed ignorance of both the IBP and of U.S. plans for participation in it. However, he expressed several times his interest in cooperating with U.S. scientists in the IBP after he had examined the papers I had with me. As I left his office he started dictating a letter to the president of the Brazilian National Research Council suggesting that a meeting be held to form a national committee for IBP.

From Belém I flew in early morning to Manaus. We made one stop en route up the Amazon. It is required that all passengers leave the plane during refueling, so during this stop the passengers stood and smoked in the shade of the plane's wings until refueling was complete. In Manaus I was to make contact with Djalma Baptista, Director, and William Rodriguez of the Instituto Nacional de Pesquisas de Amazonia (INPA). This is a large research organization for Amazonian study supported in part by the Brazilian National Research Council and in part by an organization called the "Superintendency of Economic Valorization Plan of Amazonia (SPVEA)." I called and made a late afternoon appointment with Baptista. In my hotel I discovered that I was not the only Yankee in the middle Amazon when Bill Drew of Michigan State knocked on my door. He went with me to talk with Baptista and Rodriguez and later joined me in pounding on the desk in order to get me on the only plane out of Manaus to Rio de Janeiro for the next two days after the airline denied any knowledge of my confirmed reservation.

Like Albuquerque, the two men gave us a cordial welcome and no more. We discussed IBP with them for about an hour. They expressed great ignorance about IBP but expressed an apparently sincere desire for information. They indicated their interest in cooperative research and mentioned the Amazon cruise of the research ship Alpha Helix as an example of such cooperation.

The following morning I took a half-day launch trip on the Amazon, where I went through the rites of catching a small piranha, using "dried" fish from a huge oozy slab of smelly fish on sale at a small floating backwater store. After photographing the confluence of the muddy Rio Amazonas and the blackwater Rio Negro, then experiencing a driving tropical rainstorm, I returned to the city and flew on to Rio de Janeiro.

My stay in Rio de Janeiro was a very busy one as there were many people to contact. My schedule was arranged by Dr. Bertha Lutz in whose home I stayed.

My first visit was in the office of Sr. Guido Pabst, a well-known amateur taxonomic orchidologist who was then Vice-President and General Manager of Varig Airlines, and who has a private herbarium (Herbarium Bradeanum). At the same conference was Dr. Alberto Castellanos, a botanist from the Herbarium Bradeanum and Dr. Harold Strang, former director of a conservation station for the state. The latter gave me a sizeable collection of conservation pamphlets published under his direction and talked about conservation and the needs for it in Brazil. All three of these men indicated interest in and support for IBP in Brazil.

I next visited the Museu Nacional, where I went to a called meeting in the office of the director of the museum to discuss IBP with members of the museum staff. Only three people showed up for the meeting in addition to the director. My general impression of the museum was very negative. There was no staff morale and no fire. Many people were absent from the museum although it was a work day.

The potentially most important session I had was the nearly two hours of talks with: (1) Prof. Dr. Antonio Moreira Couciero, Presidente, Conselho Nacionale de Pesquisas (CNP); (2) Dr. Hector Grillo, Vice-Presidente, CNP; (3) Prof. Dr. Manoel Frata-Moreira, Director, Scientific and Technical Division, CNP; (4) Pe. J. S. Moure, Profesor de Zoología, Universidad Nacional de Paraná at Curitiba, a bee taxonomist whom I met in Argentina in November 1966, who came to Rio specifically for this meeting because of his interest in IBP; and (5) Dra. Bertha Lutz.

We had a very amiable conversation which lasted much longer than I had anticipated. Brazil was said to be moving into participation in IBP. There was to be a meeting on January 30 to organize a committee for PM. Dr. Couciero asked Dr. Frata-Moreira to proceed with setting up a Brazilian program for IBP. I volunteered to request that a copy of the U.S. program be sent as soon as possible and this offer was eagerly accepted.

Bertha Lutz had talked earlier with Couciero about a cooperative program with the U.S. and other countries for studying amphibian genera that are intercontinental in distribution. The CNP people seemed favorable to this and asked me to transmit to Dr. Marta Vannucci in São Paulo (where I was to go next) points pertinent to this project to be included in the Brazilian program. This is the project initially discussed and endorsed by ALAIH at the Tucumán, Argentina, Congress in November 1966.

The CNP people made no mention of difficulties with U.S. scientists or organizations, but Padre Moure raised this question with reference to the Smithsonian Institution and statements that had been made by its representatives. I made the statement that the Smithsonian Institution does not officially represent the IBP program of the United States and that it cannot speak for it. I urged several time that the people in Brazil keep in mind that the real interest of most U.S. biologists is in *cooperative research* and better liaison generally with South American scientists and that it is not our desire to exploit them.

The people at CNP were enthusiastic about some kind of Interamerican News Letter pertinent to IBP. Dr. Couciero complained about the difficulty of finding out what the people in Brazil were doing, much less what was going on elsewhere. He said that the only way he could find out anything about his own people was by receiving from the U.S. the reports submitted to U.S. agencies by Brazilian scientists who had received support from them and hence had to report to them.

We discussed the low state of technology in research in such things as systematics in South America and Brazil. Interest was very strong in exchange programs that would help develop modern systematics in Brazil.

I suggested that one answer would be to try to set up centers to be developed in the newer schools by exchange with the U.S. to train in modern techniques. The idea was to get promising young scientists and encourage development by them of centers of modern research. In many older universities there is so much stagnation and dead wood that there is little hope of developing such centers there. The same may be said more emphatically for museums, of which I got mostly a very negative impression.

We also discussed need for exchange of information about publications containing biological materials, particularly the numerous obscure series and irregular series that abound in South America. The idea was put forward that the problem could be looked at jointly by a committee of U.S. and South American biologists under IBP and an effort made to assemble a listing of all such journals and publications. Nothing ever came of this.

In São Paulo, my meetings were arranged by Dr. Crodawaldo Pavan, head of the Institute of Genetics at Universidad de São Paulo. Thanks to the influence of Th. Dobzhansky over the years, this university has become a center of excellence for genetics in South America and a token of what one person of good will can do in advancing science in the lesser developed part of the world.

After a cordial visit with Dr. Mario G. Ferri, Vice-Rector of the University and a plant ecologist, a tour through the Zoology Department, and a visit to Instituto Butuntán to see the amphibian chromosome work of Willy and Maria Beçak, I met with nine people to describe the IBP. Dr. Marta Vannucci, the only member of SCIBP from Latin America and then director, Instituto de Oceanografía, helped me describe the program. With two exceptions, the group was enthusiastic and indicated that it would push for formulation of the Brazilian program. The assistant director of the São Paulo Museum and the one other person from that institution were negative and questioned the need for the IBP. They would not say that they could see the museum participating in the IBP in any way, "as the director (Dr. Paulo E. Vanzolini) was away and it would be up to him to decide." Various people stated that Vanzolini had expressed himself negatively toward the IBP. On a later occasion, he confirmed this to me, saying that the IBP was nothing but a way for mediocre people to obtain financial support. This sounded very much like statements by some

U.S. scientists, but, fortunately, this view did not prevail in the United States.

Pavan spent one day driving me to Campinas to visit the Instituto Agronómico where there is a very large program in coffee breeding and hybridization. There, I talked about IBP to a group of nine scientists. Their main interest was in the establishment of an international type collection, and I suggested that the IBP would be a logical agency through which to get it set up, but nothing ever came of this suggestion.

After leaving the Instituto Agronómico we had an interesting visit to one of three model farms owned by Dr. Paulo Noguiera, member of a wealthy and influential family. He was Presidente of Associação para a Proteção de Flora e Fauna, a private organization which he founded. The farm was interesting because no pesticides are used. There were extensive plantings of coffee, mango orchards, various tropical fruits such as jacaranda, caju, etc., as well as citrus. I saw no evidence of disease or insect infestation. The coffee trees were loaded, and mangoes were being harvested for market. I was told that peaches are protected throughout this area of Brazil by individual bagging of the fruit rather than by pesticides. Grapes and figs are raised in large quantities in this area and are said to get only copper sulfate to protect against fungus and mites. No persistent pesticides are used according to the information I was given.

On returning to São Paulo from Campinas I spent an hour talking to Dra. Marta Vannucci at her request. She wanted advice pertinent to the setting up of the Brazilian program for IBP, since it seemed apparent that she would be deeply involved. She expressed the feeling that Brazil needs all of the subprograms of the U.S. program except "Systematics and Biogeography." She said she felt that inclusion of this subprogram would encourage domination of the Brazilian IBP by old line systematists, of which there is little else of systematics in Brazil. She thought that modern systematics could enter the program under EP and elsewhere.

Dra. Vannucci and I discussed the need for developing centers of excellence in modern systematics research in Brazil and South America and she agreed with me that the best hope lies in new and developing institutions. The museums and old departments with entrenched power structures dominated by old fashioned biologists are nearly hopeless. We also discussed relations of U.S. and Brazilian PM programs. The environment being studied by her group seemed to have much in common with the Texas coast being studied by the University of Texas group at the Institute of Marine Science at Port Aranas.

Dra. Vannucci expressed the hope that I would return to São Paulo for the CT international conference in July. She thought it would be important toward keeping the Brazilian program moving. We agreed that it would be highly desirable that someone knowledgeable about the U.S. program and about South America repeat my visit twice a year for at least the first two years of IBP, especially if the South American countries start to participate actively in the program. We thought that perhaps the U.S. National Committee needed a Latin American liaison officer.

Dra. Vannucci also expressed urgent hope that she could get at least listing by title and investigator of specific projects under the various sub-programs in the U.S. program so that the Brazilians could identify counterparts in the U.S. who might organize cooperative programs with them. I promised to explore the possibilities, but we were never able to get any programs of any significance going with the Brazilians.

I flew to Curitiba in early morning, spent the day there, and returned to São Paulo in the evening. Most of the time was spent at Universidade Federal do Paraná. My visit was organized by Padre Moure, who had returned from the CNP meeting in Rio de Janeiro.

At the University I visited laboratories and institutes in Human Genetics, Biochemistry, Biology, Zoology, Parasitology, and Anthropology. The laboratories were generally adequately equipped, mostly with Rockefeller money, and some were very well equipped. Although this was vacation season in Brazil and Argentina, various people were working in the Laboratories, and there was a general air of activity.

Pe. Moure took me to visit with the Chancellor, Dr. J. Nicolau dos Santos. He spoke no English, and I no Portuguese, so the conversations were carried on in Spanish. He indicated that his people would support IBP to the limit of their ability, but he pointed out that the University would have no money to send people to conferences. A measure of their austerity came in the midst of our talk when an aide interrupted to give the president a radiogram saying the budgets of the Federal Universities had been cut 47% for next year. After that, it was hard to get back to a discussion of IBP. The chancellor, a lawyer, was said to have the reputation of not being very friendly to biology or to the United States, but he was friendly during our discussions and promised cooperation.

After lunch with Pe. Moure we visited the Instituto de Defesa do Patrimonio Natural, a state of Paraná institute under the Secretary of Agriculture. The young director, Dr. Jayme de Loyola è Silva, showed great enthusiasm for his work. The well housed and equipped institute is something of a natural history and ecological survey oriented toward conservation.

Finally I met with 10 people at the University, mostly directors of institutes and professors of departments, and explained IBP as I had done elsewhere. Response was enthusiastic from all except the head of Forestry, who defended the destruction of the *Araucaria* forest and planting of quick growing eucalypts. He could see no reason for saving any of the *Araucaria* which I was told would be completely gone in five more years unless immediate steps were taken to save a sample area, which I suggested should be done.

My impression of the Universidad Federal do Paraná was that with its good group of mostly young people, this would be the kind of place that could be developed into a center of excellence. However, the people were feeling discouraged. The phasing out of Rockefeller support had them in a panic in view of the poor support from the national government. They complained of terribly low salaries. However, they were working, and were doing work of high quality.

After this meeting Pe. Moure drove me into the countryside to find a few scattered *Araucaria* for photographing, then took me to the airport for the return to São Paulo. One of the photographs was later used on the cover of "Man's Survival in a Changing World," the booklet that was published by US/IBP describing the U.S. program in layman's terms.

On my last half day in São Paulo, Dr. Alcides Teixeira, Director of the Instituto Botánico del Estado de São Paulo, picked me up at the hotel and gave me a tour of the Botanical Gardens and Institute before taking me to my plane. The Institute was an impressive facility with good technical equipment, good libraries, and two new buildings and another one contracted for. I saw here the most impressive indication of local support for research I have seen in South America, and it comes from the state of São Paulo, which is by far the richest and most prosperous state in Brazil. All in all, São Paulo is the outstanding center for biological research I have seen in South America. Teixeira promised to support the IBP in Brazil.

My next stop was in Montevideo, where I had to shorten my planned visit because of problems with plane reservations. I visited only the Instituto de Investigaciones Ciencias Biológicas, where I was warmly greeted by Dr. C. Estable, the director and a grand old gentleman of the old school of naturalists who are interested in everything. Here, I met for the first time Dr. Francisco Alberto Saez, a pioneer student of amphibian chromosomes and discoverer of octoploidy in the escuerzo (*Ceratophrys ornata*).

Dr. Estable called a meeting to hear about the IBP, and about 12 people participated. A national committee was later set up with Dr. Estable as chairman, but nothing ever came of it. The financial situation seemed mainly responsible. I was told that the monthly income of the people in the institute was equivalent to about $100.00 (U.S.). These people quite frankly expressed hope that the IBP would mean increased research funds for them, but this was a vain hope.

In Buenos Aires I was met by Marcos Freiberg who was able to arrange meetings with key people in Buenos Aires and La Plata in spite of the fact that January is vacation month in Argentina and is taken seriously as such. In talks with Ing. Agr. Arturo Burkart, a top Argentine systematic botanist, it was obvious that Argentina was ahead of any other Latin

American country in planning for participation in the IBP. Burkart introduced me to Dr. Juan Hunziker, a plant geneticist who later became an active participant in the cooperative study of the Desert Shrub Ecosystem promoted by the US/IBP. Hunziker took us to the Faculty of Exact and Natural Sciences and showed us where, in the summer of 1966, the military invaded and lined up the faculty members before a wall and made them think they were going to be executed.

Before leaving Buenos Aires I renewed acquaintances with Dr. Bernardo Houssay, Presidente, Consejo Nacional de Investigaciones Cientificas y Tecnicas. He expressed support for the IBP. Among other things, he said that it was the moral obligation of the advanced countries to help the backward ones (like Argentina). He also said that we need more personal contacts between countries, such as those furnished by my visit and complained that I was staying too short a time.

After brief stops at the Instituto Miguel Lillo in San Miguel de Tucumán and Universidad de Córdoba in Córdoba, where very few contacts could be made because of the holidays, I flew on to Mendoza in the Argentine desert. There I met with Dr. José M. Cei, with whom I had collaborated in research for several years and with Ing. Agr. Virgilio Roig, Director of State Reserves for the state of Mendoza and later to become executive director of the Instituto Argentino de Investigaciones de las Zonas Aridas when it was established. A day in the desert near Mendoza with Cei and Roig convinced me of the close similarity in vegetation between the Argentine desert and our own Sonoran and Chihuahuan deserts.

In Santiago, Chile, Dr. Danko Brncic, a geneticist whom I had known for several years, put me in contact with a very considerable number of people at the National University. At the Estación Experimental Agronómica, I met Dr. Jochen Kummerow, a German ecologist and physiologist, who later became an active participant in the Mediterranean Scrub project. I spent one morning visiting the institute of Dr. Francesco di Castri, who later became co-director with Hal Mooney of the Mediterranean Scrub project, and who later moved to Paris to head UNESCO's Man and the Biosphere (MAB) program. Dr. Gustavo Hoeker, Dean of the Faculty of Sciences, over pisco sours in the Hotel Crillon, said that he would take the initiative in getting an IBP committee set up in Chile. A committee was later set up and a program drafted, but because of lack of funding, there was no Chilean IBP other than the cooperative effort with the United States.

I arrived in Lima at 2:00 a.m. and got a little sleep before going at 9:00 a.m. to the Servicio Nacional Forestal y de Caza, which was to be the focal point for my visit in Lima. The Director, Dr. Flavio Bazan, made me welcome although they had not expected me until afternoon. Bazan had contacted various potentially interested groups, and in late afternoon I talked about IBP to about 20 enthusiastic persons. I got an interesting introduction to Latin American custom when, as I was starting my talk, a waiter came in and gave each of those in the audience a glass of Scotch and put a glass on the speaker's table in front of me.

Bazan put a staffer and car at my disposal to visit various institutions in Lima. We went first to La Universidad Agraria la Molina, where I met the Dean of the Faculty of Science and several enthusiastic young biologists and foresters. We visited El Museo de Historia Natural of San Marcos University and talked with the Director, Dr. Ramon Ferreyra Huerta, who had been unable to attend my talk about the IBP. Ferreyra stressed the need for conservation in Latin America.

I had a very interesting discussion at the Facultad de Medicina, Universidad Peruana, with Dean Alberto Hurtado, a well-known specialist in man's adaptation to high altitudes. We talked at length about problems of science in South America. He suggested that there was a need for more people to go from the United States to South America rather than for an increase in number going from South America for training. He commented on the harm to South American science in having young people go to North America or Europe for training, where they find sophisticated hardware for research. Then they are disillusioned when they return to the meager facilities in South America, where they are disgruntled or emigrate to the United States or other developed countries. He liked my idea of trying to develop centers of excellence in special areas around young, eager, and intelligent men in provincial universities rather than trying to influence the entrenched power structures in many of the larger schools. He also expressed strong opinion that aid from foreign countries (e.g., U.S.) should go to private as well as national universities because the communist and radical student element essentially does not exist in the private universities. He gave the opinion that the communists are building for the future in trying to capture the student group in Latin America and thinks the real trouble will come in 20 years.

Bazan described a field station (Iparia) that his agency had established in the Peruvian Amazon at the confluence of Rio Pachitea and the Rio Ucayali on a forest reserve of some 250,000 hectares. I said that I hoped to visit it on some future trip as a possible IBP research site. Bazan also described an important project of his agency to save the Vicuña, which has become endangered because of shooting for fur and skins.

An IBP committee was later formed in Peru, but in spite of the enthusiasm expressed during my visit, it was never able to produce a program. As in Argentina and Chile, the only IBP research in Peru was organized primarily by U.S. scientists with participation by local scientists. In this case it was the Man in the Andes project, organized under the Human Adaptability component of US/IBP.

My last stop was in Bogotá, where I arrived considerably exhausted. Dr. Alvaro Fernandez-Perez, the botanist with whom I had made contact, had arranged a meeting at 10:00 a.m., and I arrived at El Instituto de las Ciencias on the campus of La Universidad Nacional at 10:00 a.m. Some 35 people showed up for the meeting, but all were from Universidad Nacional. Someone had neglected to inform potentially interested people at Universidad Javeriana and Universidad de los Andes. During the meeting I learned of the large national park that includes the disjunct Sierra Macarena, east of the Andes, and virtually "tierra incognita" biologically, and the though occurred that this might be a potential IBP site.

A Colombian national committee was later established, but no program ever emerged. In reverse thinking from the way things usually work, the chairman told me on a later occasion that no program was developed because no money had been made available for the IBP by the Colombian government.

As suggested by Marta Vannucci, I returned to Rio de Janeiro in July 1967 for the annual meeting of the Sociedad Brasiliera para o Progreso da Ciência. The intent was to discuss the IBP with a national gathering of Brazilian scientists and secondly I wanted to talk further with Dr. Couciero concerning the proposed symposium we had discussed during my January visit. An afternoon session was devoted to discussion of the IBP. Barton Worthington and Max Nicholson were there from SCIBP as were Ray Fosberg and Lee Talbot of the United States representing the CT section of the IBP. I mentioned the large hemispheric IBP planning conference we were expecting to hold in Caracas in November.

A meeting with Couciero was arranged by Bertha Lutz, and she, Dr. Luis Vizotto (another Brazilian amphibianologist), and I spent about one-half hour with him. With Couciero we discussed plans for a conference on modern methods of amphibian research in order to coordinate activities in the intercontinental cooperative program we had proposed in the January meeting. A tentative date was set for late February 1968, and it was my impression that Couciero would provide CNP financial support. The results were to be published in English, Spanish, and Portuguese. Subsequent efforts to get a firm commitment were unsuccessful, and the conference was ultimately held in Bogotá and without Brazilian financial help.

During my stay in Rio de Janeiro, Bertha Lutz arranged at the Museu Nacional a lunch meeting with the Brazilian amphibianologists and later a general meeting on IBP for any museum personnel who might be interested. The light turn-out for the latter meeting was prophetic of the lack of enthusiasm for IBP in Brazil.

En route back to the United States I stopped in Lima for a visit to the Iparia field station of the Servicio Forestal y de Caza. This relatively new, and very remote, station at the confluence of the Rio Ucayali and

Rio Pachitea in the Peruvian Amazon seemed to hold promise as a tropical rainforest station for IBP research. With Dr. Hugghins, a visiting Fulbright scholar, and William Guerra from the Servicio we flew by commercial DC-3 to a gravel strip at Tournavista then went by motor launch to the station. The station was well equipped, including running water and electricity until 10:00 p.m. Dr. and Mrs. James Bogart later spent four very successful months working on the amphibians there as a part of our effort to get the cooperative amphibian study under way.

The Research Planning Conference convened at Caracas on November 22-24, 1967, representing the most ambitious effort on the part of US/IBP to involve the Latin American countries in the IBP. With financial support from USNC/IBP, a group of more than 100 scientists was brought together from seven South American countries (Argentina, Brazil, Chile, Colombia, Peru, Uruguay, and Venezuela), Mexico, the U.S.A., and Australia. Forty-six institutions were represented, with the largest group coming from Venezuelan institutions and made up preponderantly of Argentine biologists who had left their country after the 1966 military coup.

The conference was organized to consider specifically two projects that had been generated in the Environmental Physiology Subcommittee of USNC/IBP. One was my project on "Convergent and Divergent Evolution," which was later modified and renamed "Origin and Structure of Ecosystems." The other project was "Physiology of Colonizing Species," convened by Dr. Calvin McMillan.

The local sponsors of the conference were Escuela de Biologia, Facultad de Ciencias, Universidad Central Venezolana as well as Asociación Venezolana para el Avance de la Ciencia (ASOVAC) and Comité Venezolano del Programa Biológico Internacional. The plenary session of the first day was held at ASOVAC. When we went to the University for work sessions on the second morning we were greeted by a hand-lettered placard denouncing two Argentine participants who had remained at the University of Buenos Aires after the military take-over in 1966. A mimeographed manifesto protesting the participation of the two Argentines in the conference was subsequently circulated under signature of 35 Argentine scientists living in Venezuela. In spite of its unpleasantness, this confrontation did not seem to affect the outcome of the conference.

After a day and a half of work sessions and a final plenary session the conference produced rather loosely defined plans for development of the two projects. Subsequent efforts to get the "Physiology of Colonizing Species" off the ground were unsuccessful. The other project was more successful, but it still required three years of program evolution and additional meetings before adequate teams of workers were actively involved in research in the field.

A chronicle of the activities that were undertaken in the two and a half years between the Caracas conference and activation of field work on the Structure of Ecosystems component in mid-1970 illustrates the difficulties of getting a complex project such as this organized and financed.

Even before the Caracas conference I was able to get a small grant to start a pilot project, with funding from the Life Science Directorate of the Air Force Office of Scientific Research (AFOSR). Harvey Savely, then director of that program, supported our efforts on the basis of his belief that prevention of wars is a significant function of the military. He felt that improvement of the level of environmental science in Latin America represented a step in this direction. With this initial support, it was possible to put three people into the field in South America. In mid-1967, James P. Bogart, on first a pre-doctoral and later a post-doctoral appointment, began field work on amphibian divergent evolution, emphasizing chromosomal studies. He and his wife drove to Peru, doing field collecting en route, and then spent some four months at the Iparia station. A Peruvian student was hired to assist the work at Iparia, partly in hope of getting him interested in the type of research being done, but this was unsuccessful. The Bogarts later did extensive field work in Argentina and Brazil between September 1969 and January 1970. In the course of their South American work they were able to obtain karyotypes of more than 160 species of anuran amphibians.

The second person was Richard D. Sage, a graduate student who in mid-1967 began field work in the Argentine Monte desert on the convergent evolution of lizards. He later continued under the NSF projects and completed the Ph.D. in 1973. The third person was William S. Birkhead, on a post-doctoral appointment, who, beginning in mid-1968, spent 16 months at Córdoba, Argentina working on convergent evolution of fishes. A young Brazilian amphibianologist was brought to the University of Texas for two months in early 1969 to learn techniques in expectation of his participation in the divergent evolution of anurans project.

The conference on amphibian research originally scheduled for Rio de Janeiro was ultimately held in Bogotá in November 1968 under local sponsorship of the Instituto Nacional de las Ciencias Naturales at La Universidad Nacional de Colombia. Nine South Americans from Argentina, Brazil, Chile, Peru, and Uruguay, 12 U.S. scientists, and several Colombian biologists participated. Alessandro Morescalchi of Italy took part and then, like several other participants, went on to Caracas for the III Congreso Latinoamericano de Zoología. The symposium papers were ultimately published in *Caldasia* (Vol. XI, No. 52) in February 1973. Because of our small grant support from AFOSR, it was possible for the U.S. Air Force to provide plane transportation to the conference for several of the South American participants, using its regularly scheduled South American flights.

Immediately following the Caracas conference, efforts were started to obtain financial support for the Convergent and Divergent Evolution program from two additional sources. In January 1968, a proposal was submitted to the Ford Foundation for establishment of four centers of excellence for this research at São Paulo, Brazil, Mendoza, Argentina, Bogotá, Colombia, and at one other location to be decided later. The idea was to provide modern research facilities at each of these and to support a mix of Latin American and U.S. graduate students and post-doctoral researchers at each. This proposal failed to impress higher echelons of the foundation and it never received support.

The second, and main, effort was made in April 1968 through a proposal to the National Science Foundation. The intent of this proposal was to represent the research interests in the convergent and divergent evolution program expressed at Caracas and to provide a coordinator for the program. Viewed in retrospect, this proposed IBP integrated research program was too diverse to ever succeed as a coordinated effort. After some months of consultation with NSF, the budget and scope were drastically reduced to four grants: one to the University of Texas for coordination and for the amphibian research under my direction, one to Arizona State University for M. J. Fouquette's work with neotropical treefrogs, one to the University of Michigan for collaborative work on convergent and divergent evolution of plants by Otto Solbrig in conjunction with Argentine botanists, and a small grant to Rockefeller University for Th. Dobzhansky's work with South American *Drosophila*.

The NSF grants were activated in early 1969. A coordinator, Gary Schnell, was employed at Austin. A post-doctoral researcher, Richard Newcomer, was sent to Bogotá, Colombia, to set up an amphibian research center at Universidad Javeriana in collaboration with interested biologists there. William F. Pyburn, of University of Texas at Arlington, led an expedition for amphibian work into the virtually unexplored Sierra Macarena of Colombia in the summer of 1969 and another in the summer of 1971. In mid-1969, NASA was able to provide a remote sensing flight across northern Argentina in support of the developing plant project. A Chilean amphibian biologist, Alberto Veloso of the University of Valparaiso, was brought to Austin for two months in early 1970 to learn modern techniques for research.

The projects centered in the three other campuses were also active in this period. Nevertheless, the program continued to draw criticism as being too loosely focused and coordinated, which it was. Consequently, a conference was convened in Medellín, Colombia, in January 1970, with encouragement from NSF in order to tighten up the project. This conference marked a significant turning point in the whole bi-continental project. The integrated research program was renamed the "Origin and Structure of Ecosystems," and this renaming was later approved by the

USNC/IBP. The total program, as restructured at Medellín, then consisted of three parts: Hawaii, Terrestrial Ecology Subprogram, added earlier, largely an independent unit and later renamed the "Island Ecosystems Subprogram," an "Evolutionary and Ecological Diversity Subprogram" for the divergent evolution studies of amphibians, and the "Structure of Ecosystems Subprogram" for the convergent evolution studies. Otto Solbrig was named director of the third of these subprograms. This subprogram was split into two separate projects which became the "Desert Scrub" project co-directed by Solbrig and by Jorge Morello of Argentina's Instituto Nacional de Tecnologia Agropecuaria (INTA) and the "Mediterranean Scrub" project co-directed by Harold Mooney of Stanford and Francesco di Castri of Universidad Austral in Valdivia, Chile.

Following the Medellín conference, two separate conferences were held early in 1970. The Structure of Ecosystems group met in San Diego, California, to plan the integrated project that was finally funded in the fall of 1970. The Ecological and Evolutionary Diversity group met in Lawrence, Kansas, but for various reasons, it was decided not to seek support for this work. From this point on, the South American-North American component of the Origin and Structure of Ecosystems IRP was focused on the comparative study of two sets of apparently convergent ecosystems, the Mediterranean Scrub systems of southern California, and of Chile on the one hand and the desert scrub systems of the Sonoran Desert and the Argentine Monte on the other.

In July 1970, Otto Solbrig, Charles Lowe, and I made a survey of sites in the northern Monte and tentatively settled on Cafayate in the valley of the Rio Santa Maria. A research planning conference was held in Argentina in December, with sessions in Tucumán, Cafayate, and Salta. Twenty-three U.S., 22 Argentine, and one Venezuelan scientist participated. Subsequent to this conference, it was decided to move the central study site to Andalgalá in the state of Catamarca, as that area provided a better match for the northern site near Tucson than did Cafayate.

In October 1970, Harold Mooney and Albert Johnson visited Chile and in conjunction with Francesco di Castri and Ernest Hajek selected a Mediterranean Scrub site at the Fundo Santa Laura on the Cuesta la Dormida, north of Santiago. A research planning trip for Mediterranean Scrub participants, comparable to the earlier Argentine one, was held in late April-early May 1971, and terminating in Valdivia.

Thus, nearly three years after the Caracas conference the Origin and Structure of Ecosystems IRP was finally shaped up into a clearly focused interamerican research effort.

In the period from early 1968 onward I made every effort I could to promote the IBP in particular and modern biology in general in Latin America. Several extended trips through South America were made in this period, and many contacts with Latin American biologists were made or renewed. One of our continuing and futile efforts was directed toward organizing a Tropical Forest Biome program comparable to the biome projects of the US/IBP but on a cooperative basis with Latin American scientists.

In March 1968, I went to Santa Fe, Argentina, to participate in a regional IBP meeting organized by the London office for the PF section and sponsored locally by Dr. Argentino Bonetto, then Director, Instituto Nacional de Limnología. En route, I stopped for consultation in Bogotá and Cali, Colombia, Lima, Peru, and Mendoza, Argentina, and had an airport meeting in Santiago with Francesco di Castri and other Chilean scientists.

My principal objective in Bogotá was to meet with a group from several Colombian universities and to go with them to discuss with Dr. Bateman, Colombian minister of Public Works, the disposition of the so-called Battelle camps on the Rio Atrato. These were built by the United States as bases for the feasibility study for a sea-level canal, and they would be turned over to the Colombian government on termination of the study. It had occurred to several people that these might serve as tropical field stations for Colombian universities and that a tropical ecosystem program under the IBP might be developed around them. In Bogotá I met Alvaro Fernandez-Perez, Jesús Idrobo, and Jorge Hernandez from Universidad Nacional de Bogotá, Hubert Hoenigsberg from Universidad de los Andes, Jaime George from Universidad Javeriana, and Fabio Heredia-Cano from Universidad de Antioquia. We had a very amiable meeting with Bateman at which it was agreed that:

(1) The National University in Bogotá would immediately initiate a request for the two camps nearest the coast. This would have to clear through university officials, but no problems were anticipated.

(2) A consortium of universities would be set up, presumably with recommendations from the Colombian Society of Naturalists, to manage the camps. It was felt that a lease arrangement at token charge could be effected to put the camps under operation of the consortium.

I also recommended a third step, which would involve a small conference at the camps between representatives of participating universities, three or four interested persons from the U.S., and an OTS representative. The purpose of the conference would be to plan a tropical ecosystem program.

The purpose of my stop in Cali was to make contacts with biologists at the Universidad del Valle and also to meet Carlos Lehmann, Director of the Natural History Museum. Ignacio Borrero, Head of the Biology Department, saw to it that I met faculty and administrators and had an opportunity to explain the IBP to them. As usual, I stressed the integrated programs for ecosystem studies, especially for the study of the tropical rainforest and indicated ways in which these interests relate to the U.S. effort under IBP.

One positive result of my visit to the Biology Department was that one young biologist, Reinaldo Diaz Vergara, later came to the United States to do a Master's degree with me.

The meeting with Carlos Lehmann was extremely interesting. I had known him only by his reputation as probably the most enthusiastic and effective conservationist in Latin America. At the University, I was told that the four-year-old museum which he headed was the third museum that he had founded in Colombia.

Lehmann took Borrero and myself for a 75 km ride up the Cauca Valley to a fishery research station that had been given to the university. The native forest is gone except for tiny clumps and isolated trees and large clumps of a very large native bamboo (*Gauda*). The latter remains because this bamboo is used so much (scaffolding for building, safety poles at RR crossings, etc.) that it is forbidden by law to clear it away completely. Lehmann did most of the talking while dodging wildly driven trucks and while drifting nearly off the road as he leaned over to make a point. Many of the points he made are worth recording:

(1) He said that the greatest threat to conservation in Colombia was the export of wildlife from Leticia on the Amazon. "More than 20,000 birds have been sent to New York in one shipment. The caiman has been eliminated from the area of the Amazon covered by the collectors for this trade."

(2) Birds being shipped to the U.S. are in many cases species that migrate between North and South America, so we are protecting them at home yet destroying the same population by permitting their import. For every bird that arrives alive in New York, 19 have died after capture! In some types (e.g., cock-of-the-rock) the ratio is 1 to 50!

(3) The U.S. government sends advisors to high-ranking people in South American governments, but they never send people competent in ecology. Instead, they send engineers to advise how to destroy what is left of South America. "We have enough of our own poor engineers doing this, so we don't need any help from poor North American engineers." On the other hand, ecological advice at high governmental level would be of great help toward better conservation practices in South America.

(4) Peace Corps workers should have ecological indoctrination before being sent into the field. They are indoctrinated in many other ways. Ecological indoctrination would make them a force for development of a concept of conservation in the countries in which they work.

(5) Translations of such biology texts as the BSCS Series and Odum's Ecology are very poor. The words are Spanish, but the constructions are English.

(6) Most of the extinctions of native animals (over 50 among Colombian birds alone) have resulted from the destruction of their habitats.

(7) Streams in western Colombia are badly polluted. The pulp industry and sugar cane mills are the most sources. When effluents from paper mills are dumped at regular intervals, the natives gather for many kilometers downstream to pick up the dead and dying fish.

(8) Pesticides have been and are being badly used by people who cannot understand the dangers of incorrect dosages. There have been deaths. Tomatoes, which are usually eaten without being washed, are often heavily dosed, with resultant illness. In one instance, airplane spraying was done without attention to wind direction with the result that "kilometers of river" were white with floating, dead fish.

In one instance insecticide was sprayed over an agricultural crop to kill mourning doves which were concentrating there. The intoxicated doves were shot by hunters who ate them and subsequently were hospitalized.

(9) Introduction of trout has done untold damage to native biotas and has resulted in extinction of endemic species in some known instances. Trout were introduced in Lake Tota and the lake was closed to fishing. The trout destroyed an interesting endemic fish (*Rhizosomichenys totae*) and then became cannibals. Since trout have been introduced into many places in the Andes, one can only guess at how often this has happened.

(10) Engineers built a hydroelectric dam in the Cauca Valley, and because they were afraid that the dead leaves from the trees of the surrounding forest would clog their turbines, they cut down the forest. Now, the adjacent slopes are slumping into the lake.

(11) Certain species of *Eucalyptus* have been introduced in the Cauca Valley to "dry up" the marshy areas so they can be used for sugar cane. This succeeded so well that the water table dropped to where "wells had been drilled for irrigation, and some of these have become dry."

(12) In some areas the native vegetation has been destroyed and replaced by pine in pure stands. Scarcely any of the native birds or other fauna enters the pine forest. This year there was a damaging outbreak of a beetle in the pine forest that was so bad that "workers were removing them by shovelfuls."

(13) Destruction of marshes has resulted in decline of the native herons and egrets and has not affected the invading cattle egret, which is very abundant.

Lehmann, Borrero, and I talked at some length about the Conservation of Ecosystems (CE) program in the Americas. Lehmann had had considerable interchange with SCIBP, but none with the U.S. program. The outcome of our discussion was a consensus that cooperation was needed among CE committees of the Americas and that the most useful thing to do would be to pool their efforts to produce a catalogue of the most important areas that are threatened in the Americas and then to publish this in the form of a well written, well illustrated, attractive book that would have appeal to the general public. Criteria for inclusion would be uniqueness, representation of vanishing biomes, presence of rare and endangered species. Unfortunately, we were never able to follow up on this.

I stopped in Lima to renew contacts at the Servicio Forestal y de la Caza and with academic scientists. At the Servicio, Flavio Bazan, the director, told me that he was leaving Peru to head an FAO forestry project in Nicaragua. Bazan was chairman of the only IBP subcommittee in Peru. This was the Conservation of Ecosystems (CE) committee, and Bazan called a meeting while I was there. Ian Grimwood of IUCN was in Lima and also participated. I urged that additional subcommittees be formed. I also presented the idea of an illustrated catalogue of the endangered ecosystems of the Americas. There was no dissent and little discussion. Hans Koepcke presented the point that there was inadequate information about animals of Amazonas that he as an ecologist needed.

At the Museo de Historia Natural Dr. Ramon Ferreyra Huerta, the Director, called a meeting in his office to discuss the IBP. Present were Ferreyra, Dr. Luis Gonzales-Mugburu, President of the Peruvian Association of Biologists and chairman of the Peruvian Committee for IBP, Dra. Luz Sarmiento, Dra. Nelly Espinoza, Dr. Hans Koepcke, Dr. Oswaldo Meneses Garcia, Instituto Nacional de Salud, a parasitologist who was presently to make a survey of the venomous animals of Iparia, Dr. Pedro G. Aquilar F., Universidad of Agraria la Molina, an entomologist interested in field crickets.

A result of this conference was a consensus that a cooperative IBP program could be organized involving several subcommittees and integrated programs. It was agreed that an effort of integrated nature centered at El Parque Forestal Nacional de Iparia would be a logical enterprise. Dr. Williams at the Field Museum was already studying the plants. James Bogart, my graduate student, had spent four months studying the amphibians. Dr. Oswaldo Moneses Garcia would soon begin a study of the venomous animals, with support from AFOSR of the U.S. I agreed to send students to study other groups, e.g., the small mammals. The Koepckes were moving soon to the area of Iparia for ecological studies. Hence, it seemed possible to "beef up" this program to the extent that it could become an ecosystem study program. We were never able to get support for this proposed cooperative project. Señora Koepcke was later killed in the crash of a commercial plane in the Peruvian Amazon, and the daughter received world wide attention as the only survivor because of her survival in the tropical forest for several days before reaching help.

Between flights in Santiago I had an hour to talk with Francesco di Castri concerning the progress of the IBP in Chile. I stopped for a day in Mendoza to check on the progress of my student Sage, then flew on to Santa Fe.

When I arrived at the small Santa Fe airport there was no one to meet me. My letter of ten days ago had not arrived. Not knowing the name of my hotel I went by asthmatic bus to the Aerolineas Argentinas terminal in Santa Fe. I had no pesos and could not obtain "cambio" at the small airport, but a friendly Argentinean paid my 150 peso bus fare and stood by to see that I was taken care of by Aerolineas Argentinas. They called Bonetto's office and also called a taxi and changed $1 to pesos so I would have taxi fare. They also told me how much taxi fare to pay (about 30 cents).

The meeting in Santa Fe had a good representation from Latin American countries. There were people there from San Salvador, Colombia, Peru, Chile, Argentina, Brazil, Uruguay, Bolivia, and Dra. McConnell of the UK spoke on behalf of British Guiana. Franz Weibezahn from Venezuela did not come, although he had been sent a ticket. It seems that if he had come, the refugee Argentineans would have demonstrated at Caracas and would have caused him trouble! Marta Vanucci from São Paulo also backed out at the last moment, although Col. Krause had arranged Air Force transport for her. It was not clear whether this had political significance or not. Art Hasler and I and Colonel Richard Krause of USAF from Rio de Janeiro embassy were the norteamericanos.

The meetings were very formal. We went by car and taxi to the Cámara where we were to meet the president of Santa Fe State and have our first session. However, after picking up our papers, we bustled off to the President's house (Palace) and were received briefly by him. Just as we were leaving there was a tremendous thunderstorm. Eventually we left, getting well soaked and returned to the Cámara. The driver of the taxi I was in failed to turn on windshield wipers, which made me a bit uneasy, but we arrived back at the Cámara. There was a very formal opening session, with the President presiding, but after this he and his dignitaries departed and discussions by Bonetto were begun and responded to.

The meeting was unlike any North American meeting I had ever participated in. Each delegation sat together with labels and with national flags in the background. Teams of two shorthand reporters replaced one another every few minutes and the meeting was also being taped. There was much talk of pisciculture and of shortages of men, money, and equipment, but no concrete planning was done. I was convinced that the greatest value of the meeting would be the establishment of contacts among the people of the various countries.

During the session I spoke about the U.S. program and its development into large ecosystem studies. I also mentioned the five integrated U.S. programs that related considerably to Latin American participation: (1) convergent and divergent evolution, later renamed "Origin and Structure of Ecosystems," (2) Physiology of Colonizing Species, (3) Experimental, Analytical Biogeography of the Sea, (4) Human Adaptability, and (5) Conservation of Ecosystems. The second and third of these never got off the ground in the United States. I also mentioned the need for cooperation across national boundaries in Latin America in developing greater competence in biological science. I also mentioned the IBP Interamerican Newsletter.

After leaving Santa Fe I missed Dr. Avelino Barrio in Buenos Aires, but dined with Dr. Marcos Freiberg, then flew to Rio de Janeiro the next day. There I discussed amphibian biology with Dra. Bertha Lutz, Dr. Luis Dino Vizotto of San Jose do Rio Preto and Dr. Eugenio Izecksohn of Universidad Rural.

My next trip to South America came in June 1968, and again a main object was disposition of the Battelle camps and the promotion of a tropical program. Webb McBryde from the Battelle organization in Panama and Steve Preston from the University of Michigan, representing the Organization for Tropical Studies, joined me in Bogotá. We went with the same group of Colombians, representing four universities, for another talk with Dr. Bateman.

Dr. Bateman agreed after considerable discussion that the biologists could have the use of the "Alto Curiche" and "Teresita" camps. This seemed a highly satisfactory arrangement. Teresita was scheduled to be turned over no later than January to April of 1969 and Alto Curiche no later than April of 1969. Dr. Bateman indicated that he thought the AID people might strip the camps before turning them over. I agreed to check on this in Washington.

Webb McBryde said that Colonel Sutton in command of the Canal survey would welcome work at the camps before the turnover. Preston and

I made plans to stop to talk with Colonel Sutton in Panama. McBryde also indicated that there would be camps available under the same circumstances in Panama. The consensus during and after the meeting was that it was most important to get the biologist's "foot in the door" by getting as much work under way at the camps as soon as possible.

After this we went to a meeting of the Colombian National Committee for the IBP. They had sent out a questionnaire aimed at making a directory of Colombian biologists. The most important decision made in this meeting was that the Committee apply to the Organization of American States for a grant under the "Acciones Refuerzos" program on behalf of all Colombian universities to develop an international center for tropical biology, using the Chocó camps and La Reserva la Macarena as their chief assets. Preston told the Colombians about OTS and indicated that a consortium of Colombian universities might join, using these facilities to fulfill their obligations to the OTS consortium.

From Bogotá, Steve Preston and I flew with a group from La Universidad Nacional de Bogotá and La Universidad Javeriana and from INDERENA to examine La Reserva la Macarena as a possible IBP research site. We failed to reach the mountain because high stream flow hindered our progress by launch upstream, but we saw some excellent lowland rainforest and got a sampling of the kind of subsistence agricultural existence lived by the human population of the rainforest.

We took off from the military airport at Bogotá in early morning in an old C-47. First we flew to an airbase in the Magdalena Valley to gas up as the pipe line to Bogotá was delivering contaminated fuel. Then we flew to San José de Guivare, a tiny river town where we lunched on food "criolla" in a mud-floored "restaurant." My appetite would have been better if I had not been seated where I could see back into the lean-to kitchen, where a young javelina was rooting about on the mud floor and a spider monkey was hopping about. The waitress, who had some ugly sores on her face, would place a piece of fried beef on a plate, lick her fingers and then repeat the process. After lunch we and the supplies were ferried in segments to an INCORA camp at the junction of the Rio Ariari and Rio Guayabero. A few difficulties were encountered, however, and the last segment of the party ran out of gas and had to stay overnight at a "house" about eight miles downstream. The food was loaded on a river boat which did not arrive until 10:00 a.m. the following day. At about 9:00 p.m. we had "comidas criollas" served by a typical family living near the station. Their establishment had to be reached by dugout canoe across a flooded creek. We had *yuca, plantanos, papas,* and *arroz* in a boiled dish and the toughest boiled chicken I ever ate.

The next day's breakfast was taken at the same place, with fried *platanos* and boiled rice and tomato ketchup to spice it up. There was also cacao, which tasted fairly good.

This family was living in a low area adjacent to the river with a planting of rice, corn, and *yuca*, along with sugar cane. Coffee and cacao are available across the creek. There was a scraggly *Opuntia* said to come from Mexico and treasured for its tunas.

After two days of runs up the river and being frustrated by high waters in the *angosturas* (rapids) we returned to San José to wait several hours for the Colombian Air Force plane to come for us.

Back in Bogotá, Steve Preston, Alvaro Fernandez-Perez, Jaime George, and I spent an afternoon working out the outline for a proposal to be submitted to OAS by the Colombians. This was based on the availability of the Macarena with three biological camps, the Chocó with two anticipated stations, the páramo and the llanos, with one already existent station at Villavicencio. Unfortunately, this project never materialized.

In Panama, Webb McBryde and I met with Colonel Alex G. Sutton, Jr., Field Director, Atlantic-Pacific Interoceanic Canal Study Commission. Colonel Sutton was very cooperative and said he would be glad to have IBP people at the remaining camps in the Chocó. He wanted a letter stating purpose of work, names of people, and duration of their stay. Out-of-pocket costs to the government would be charged. He wanted access to data accumulated. He also indicated willingness to turn over to Colombia a complete weather station at Alto Curiche if the Colombians would operate it. He promised logistical support when available.

We then went to talk with Dr. Quieros, Minister of Public Works and Education and a member of the Canal Commission. We discussed possible IBP use of a camp in the Darién that had reverted to a private citizen. We also discussed formation of a Panamanian National Committee for the IBP. McBryde promised to push this.

Webb McBryde hosted an evening reception at the Officers Club in the Canal Zone, and about 60 people, including wives, attended. During the evening several of the people indicated interest in a National Committee for the IBP. I promised to send information about the IBP to McBryde, who would distribute it to interested persons.

My next trip to South America was in early 1969 and took me to Bogotá, Florencia, and Villavicencio in Colombia, Córdoba and Mendoza in Argentina, and Santiago and Valdivia in Chile. In Bogotá we had a two-day meeting of IBP national chairmen from American countries. I had promoted this meeting for some time as a mechanism to help coordinate IBP activities in the hemisphere. The Instituto de Ciencias Naturales was the Bogotá host. Present were Dr. Armando Cardozo of Bolivia; Dr. N. F. Gibbons, Vice-Chairman of the Canadian National Committee; Dr. Francesco di Castri, Chairman of the Chilean National Committee; Dr. Eduardo Mora, Chairman of the Colombian National Committee;

Dr. Misael Acosta-Solis (Chairman) and Col. Marcos A. Bustamante Y. (Coordinator) of the Ecuadorian National Committee; Dr. C. Hernando de Macedo, Chairman of the newly organized Peruvian National Committee; and myself as Chairman of the U.S. National Committee.

Dr. Osvaldo Boelcke of Argentina and Dr. C. Estable of Uruguay were unable to attend because of, respectively, a conflicting meeting and doctor's orders not to travel while convalescing from illness. Dr. F. Weibezahn of Venezuela telegraphed last-minute information that he had to change his plans to attend because of a recent development. No reply to the invitation to attend was received from Brasil.

The participants described the state of progress of the IBP in their respective countries. Dr. Cardozo explained that no IBP committee existed in Bolivia but that he would form one on his return home and petition the SCIBP office for recognition of his country as an IBP participant.

Several resolutions were passed as recommendations to their respective organizations. These were:

(1) Recommend another meeting of National chairmen, or their delegates, on November 10-11, 1969, in La Paz, Bolivia.

(2) Recommend a meeting of directors of programs involving the systems approach, particularly "Ecosystems Analysis" and "Convergent Evolution" in Washington, D.C. on 12-13 September 1969, following an IBP grasslands meeting in Saskatoon.

(3) Recommend that the various committees of the American countries participating in the IBP explore ways of improving information exchange (publications, names of scientists, and projects) among the American countries.

(4) Recommend one central Information Exchange Center in Latin America, with a subsidiary center (perhaps one person) in each participant country.

(5) Recommend consideration of one center for data storage for all systems-oriented programs in the Americas, and suggest discussions with Director of Ecosystems Analysis Program of the U.S. concerning the possibility that the U.S. center be so designated.

No one of the five resolutions was ever implemented. The La Paz meeting was cancelled as the Bolivian representative had to leave his country because of a political change, and it was never rescheduled elsewhere. The general problems addressed by the last three resolutions still remain as significant challenges.

From Bogotá I bussed across the Andes to Florencia with students and faculty of the National University to take part in the II Amazon Forum to be held there and at Leticia. During these meetings, Helmut Lieth, who was then director of the on-paper Tropical Forest Biome and who was in Colombia hopefully evaluating possible IBP sites for this biome, organized a conference with the Colombian Minister of Agriculture to discuss the IBP. Meeting with the Minister were Enrique Perez-Arbalaez, Jaime George and Alvaro Fernandez-Perez of Colombia, Leith, Dick Cowan of the Smithsonian Institution, Dick Davidson, Bill Martin, and Jim Duke of Battelle, and myself. About 25 minutes were spent explaining the IBP and telling the Minister of the need for financial support. At the end he said that he could provide some money and would give his support in efforts to get other money from the government. This seemed to be a very successful meeting, but nothing ever came of it, as was unfortunately true of most of our meetings in South America.

The degree of representation of Amazonian countries in this "Amazon Forum" is illustrative of the dearth of inter-country communication and of the difficulty of participation in international conferences even within the continent of South America.

This was very nearly a Colombian event, plus strong participation from the U.S. Of the 101 registrants, 72 were from Colombia (about 40 were students from Universidad Nacional in Bogotá), 21 were from the U.S., 3 were from Ecuador (2 there because of the IBP meeting in Bogotá the previous week), and there was one each listed for Brasil, Mexico, France, Germany, and England.

The "Forum" was to proceed from Florencia to Leticia on the Amazon, but I left them to fly to Villavicencio with Fred Medem to see the Instituto of which he was director. We flew by way of Satena Airline which consists of beatup C-47 planes, operated by the Colombian Air Force to provide transportation in the virtually roadless eastern part of the country. This flight gave me a sampling of what air travel is like in the Colombian Amazon.

The bucket-seated C-47 arrived two hours late. The passengers were a remarkable mixture of grizzled colonos with their machetes and worldly belongings, a few well-dressed couples apparently returning from vacation, and three young men with cameras who looked as if they were Colombian newsmen in search of a story.

As the plane took off, a chubby, well-dressed young Señora across from us grasped her husband's arm as if it were her first flight, and she was soon using the brown paper bag although the flight did not seem overly rough for a C-47 in the tropics.

We landed at San Vicente del Cagúan, where all landing passengers had their luggage thoroughly searched by a detachment of soldiers. Medem said this happened occasionally in the Amazon country of Colombia. At San Vincente several passengers got on the plane. In addition, we took on six huge cheeses covered by gunney-sack material and each weighing in excess of 100 pounds, judged by the efforts of the two men who heaved each on board.

When we left Florencia several passengers were left behind because they were going to La Macarena, and they were told that the plane would not land at La Macarena. However, for some mysterious reason, probably freight, we did land next at La Macarena. There we took on two or three passengers, a huge catfish of probably 100 pounds or more, a large roll of dried javelina (peccary) skins, and various boxes of who knows what.

This area was one of "bandoleros" until recently when their leader was killed. To do field work in this area, according to Fred Medem, one had to obtain permission from this leader, a dangerous paranoid.

After leaving La Macarena we flew across La Sierra Macarena. This is biologically a virtually unexplored area. First we flew over large areas of scrub vegetation, with gallery forest along the streams. There is much rock outcrop. Some say this is a remnant of the Guayana Shield; others say no. Toward the east, the vegetation becomes selva. After crossing the Macarena we flew over the llanos, with scattered forest or none and broad, braided, sandy rivers.

I spent a day in Villavicencio visiting the Instituto Roberto Franco, which belongs to the Universidad Nacional in Bogotá and is directed by Medem. It is quartered in the former Rockefeller Yellow Fever Institute. The chief activity at the Institute by Medem was the breeding in captivity of various turtles and crocodilians including some rare and endangered species. Medem showed me a diversity of side-necked turtles and the only *Paleosuchus* (a rare crocodilian) to be raised in captivity.

Medem and I went by taxi across the Andes to Bogotá. From there I flew south to Córdoba, Argentina, after getting my pocket picked in the mob of people trying to board flights at the Bogotá airport. In Córdoba I checked on progress of Bill Birkhead, my post-doc working with fish, and then in Mendoza I did the same with graduate student Dick Sage, working with lizards. In Mendoza I discussed the IBP with Zoologist Ing. Vergilio Roig and his botanist brother Ing. Fidel Roig.

Going on to Santiago I joined up with Dr. Alberto Veloso of the Universidad de Valparaiso and a Santiago herpetology student, Klaus Busse, for a field trip southward to Valdivia in the southern *Nothofagus* (beech) forest region.

This proved to be one of the most interesting and most productive opportunities to do actual field work of any of my South American trips. We were shown about the region near Valdivia by Dr. Hugo Campos, Director, Instituto de Zoología, Universidad Austral. As a result of these Chilean contacts, both Veloso and Campos later came to the University of Texas for post-doctoral training. From Valdivia I flew back to the United States.

The II Jornadas Argentinas de Zoologia was scheduled for mid-September, 1969, in Santa Fe. This seemed like an excellent opportunity to publicize the work we had gotten under way on convergent evolution, so

I suggested a "Mesa Redonda" on convergent evolution to Dr. Argentino Bonetto, organizer of the meeting. I understood that this suggestion was accepted and asked Bill Birkhead to present his work on fish and Dick Sage to do the same with his work on lizards. I would do the same for amphibians. However, when we arrived in Santa Fe we found our papers scattered throughout the program with no indication that they were parts of a cohesive effort.

During the Santa Fe meeting several efforts were made toward better coordination and cooperation in IBP projects between the United States and Argentina. One dinner meeting was held involving persons interested in the Convergent and Divergent Evolution program as it then existed. Present were Gary Schnell, coordinator for the project, Avelino Barrio from Instituto de Microbiologia Malbran in Buenos Aires, José Cei from Universidad Nacional de Cuyo in Mendoza, Abraham Willink from Instituto Miguel Lillo in Tucumán, Jorge Morello from INTA, Bill Birkhead, Dick Sage, and James Bogart from my group, and Ricardo Timmermann, a student from Universidad de Córdoba.

One half-day session of the Jornadas was devoted to the IBP. Osvaldo Boelcke, chairman of the Argentine IBP Committee, described their planned program. I described the U.S. program emphasizing the IRP (Integrated Research Program) concept. I also talked about the relation of IBP to UNESCO's proposed MAB (Man and the Biosphere) program and of a proposed global network for environmental monitoring. In the discussions following our presentations, as well as in private conversations, a number of frictions and grievances between our people and the Argentines became apparent. Some of the grievances involved personality conflicts, such as the very bitter one between Cei and Sage. The biggest general problem seemed to be the need of better coordination between the two national programs for IBP. Argentine scientists who had become involved in our U.S. projects seemed unwilling to be identified with the official Argentine program, and this not unnaturally rankled with the Argentine committee. As a result of these discussions, Boelcke invited me to lunch with the big wheels of Argentine biology to discuss better coordination between the two national programs. Those present included Barrio, Cei, Bonetto, Willink, Raul Ringuelet and Ricardo Ronderos from La Plata, and one person whom I did not know.

We agreed to coordinate programs. I would send a description of the Convergent and Divergent Evolution Program to Boelcke. I would list Argentines who have indicated they wanted to participate. The Argentine National Committee would have to approve the plans of the various Argentine participants, who would be asked to submit plans and financial statements. We also arrived at a general plan for better coordination. There would be an Interamerican Commission with representatives from the participant countries. There would be subcommittees under each national IBP committee. Research within any country would be coordinated through these committees. Knowing something of the crosscurrents or Argentine biopolitics it was not surprising that nothing ever materialized from this effort at coordination.

Before leaving Argentina I visited INTA in Buenos Aires with Jorge Morello to talk with some of their top people about processing of the data from the NASA remote sensing flight across northern Argentina in July. We also discussed the potential of the earth resources satellite (ERTS) for global monitoring.

Later in 1969 I was invited to take part in the II Congreso Nacional de Biología in Lima and following that to give a one-week short course on modern mammalogy at San Marcos University under a Fulbright Lectureship. En route to Lima I made two stops in Colombia to continue the effort to get some IBP research started in the tropical rainforest. In Bogotá I met with Jaime George of Universidad Javeriana and then Dick Newcomer, who was in Colombia doing post-doctoral research on frogs under my IBP grant, and I went to INDERENA headquarters to talk with Jorge Hernandez about a possible Macarena expedition in July of 1970. He suggested a one month effort in July.

We went then to the Instituto Nacional de las Ciencias Naturales where we met with Dr. Eduardo Mora, Director and Chairman of the Colombian IBP Committee and Pedro Ruiz-C and Alvaro Fernandez-Perez of the Instituto. Mora insisted that the Instituto be the central coordinator for the Macarena expedition, which was a reflection of previous conflict between INDERENA and the Instituto. The Parque Nacional de la Macarena had been assigned to U. Nacional for research and INDERENA had administrative control.

From Bogotá I flew to Medellín, then Fabio Heredia and I flew to Quibdó in the Chocó where the annual rainfall is in the neighborhood of 400 inches. We were still looking for possible IBP sites in the rainforest. We saw the Rio Atrato by motor launch and had an additional trip into the countryside by pickup truck. We found the forest very much disturbed by timber cutting and we found Quibdó rather dirty and unattractive.

In Quibdó we met briefly with the regional head for INDERENA, and then with the governor of the Chocó state. The governor was quite friendly and offered assistance in any project the IBP might develop in the Chocó. He offered use of a house owned by the government in the city and sent an assistant with us to look at it. The house was abandoned and filthy, but it was very solidly constructed. For an estimated $2,000 it could be made into a very good headquarters for any IBP project that might be developed in the Quibdó area. However, we left Quibdó with reservations about trying to develop a program there, largely because of the disturbed nature of the forest.

Near the end of my stay in Lima, Dr. C. Hernando de Macedo, Chairman of the Peruvian IBP Committee, and I arranged a meeting with Dr. Giesecke, Director of the newly formed National Research Council, to discuss the IBP and the multinational centers program of the OAS. Macedo was to follow up on this. A possible project would be an Andean center with subcenters for investigation of the Andean biota. Possible subcenters could be Lima, Santiago, Bogotá. I promised to talk about this with Jesse Perkinson of the OAS in Washington. However, nothing ever came of this proposal.

In the course of the trip in July 1970 by Otto Solbrig, Charles Lowe, and myself to check out research sites for the Desert Scrub project in Argentina, we had several discussions with people at INTA.

The first meeting was with Angel Marzzoca, Assistant Director for Special Projects of INTA, Antonio Prego, Director, National Center of Natural Resources, and Ubaldo Garcia, INTA research director. The meeting lasted for more than two hours. INTA was interested in our program and the officials present promised to cooperate. Garcia indicated that his support would be given if there was inclusion of a strong element of training. Marzzoca asked what we needed from INTA. Items mentioned were vehicles (2) to be made available, with upkeep from U.S. participants, easement of import restrictions on research equipment, and most importantly the contribution of personnel from INTA. Prego agreed that Morello could spend one-half time for six months of the year on this program and also indicated that about three persons at Salta could be assigned to the project for work as needed.

Marzzoca asked for a "convenio" to be signed by INTA and U.S. representation. He and the others of INTA were not enthusiastic about IBP-USA as a signatory, as the IBP had a finite life, and they were thinking in longer terms. We felt that NAS/NRC through the Office of the Foreign Secretary would be the logical signatory. This would be consistent with our thinking about the follow-up on IBP through national and international SCOPE.

Before returning to Buenos Aires from our field trip, we drafted a tentative version of an "acuerdo" between INTA and U.S. NAS/NRC regarding the Desert Scrub program. A comparable document between INTA and a U.S. university was heavily plagiarized to form the body of the document. This was done in preparation for our further discussions with INTA officials.

Back in Buenos Aires we met for nearly two hours with Marzzoca, Prego, and another INTA official whom I had met on my last trip to Buenos Aires. We discussed and made some small alterations in the "acuerdo" which our group had drafted. Much of the discussion centered on the population problem (on which the Argentines were divided) and on the question of environmental degradation. We were asked why there was now so much interest in ecology in the United States. I explained that it came from concern about pollution and environmental degradation. When I asked if there were basic ecological studies being made on pampas grasslands, the INTA official replied by asking what could be the use of such, since there were no pampas grasslands left except for a few patches along roadsides. Prego said repeatedly that pollution was showing up in Argentine rivers and mentioned specifically the Rio Segundas in Córdoba.

On my return to the United States I talked with Harrison Brown, Foreign Secretary of NAS/NRC, about that organization signing the acuerdo with INTA. The answer was that the bureaucratic hurdles to such a signing made it unlikely. The upshot was that the acuerdo was eventually signed by Harvard University on behalf of the Desert Scrub Program.

In December 1970 I participated in the Desert Scrub program planning sessions in Argentina. There was a good mix of U.S. and Argentine scientists. We traveled by chartered bus from Tucumán to Cafayate to Salta with field stops in between and talks about their research plans by the participants. A group of anthropologists from La Plata participated in the discussions at Horco Mollo at Tucumán, but this group unfortunately failed to get involved later in the actual research.

In February 1971 I went to Lima as an invited participant in a forum on "Environmental Contamination." Much of the impetus for this forum was credited to Bernardo Batievsky, a wealthy Lima businessman. In addition to participating in the forum and giving a talk on "Contaminación Ambiental," I gave two public lectures. At the Museum of San Marcos University, I talked on "Evolución Convergente de Ecosístemas Secos." At Universidad Cayetano Heredia I lectured to a large group of students on "Biogeografía General de Sudamerica."

Bill Milstead and I participated in the Mediterranean Scrub planning conference in Chile in late March, 1971, and then bussed across the Andes from Puerto Montt, Chile, to San Carlos de Bariloche in Argentina. There we visited with Dr. E. H. Rapoport, a zoologist I had first met in Caracas in 1967 and Dr. Carlos Malman, Director of Fundación Bariloche. We discussed the IBP and other programs, UN, SCOPE, OAS, that might involve interamerican biology. We then went with Julio Contreras to see the biological station of the Fundación of which he was director. This station is located on Isla Victoria in Lago Nahuel Huapi, which is included in a national park.

En route by motor launch Contreras gave us some statistics about the island and the park. The lake is a very large, very deep, very cold, glacial lake. Introduced trout have virtually exterminated the three native galaxiid fishes, which were plankton feeders, so the food chain now is broken. The trout fishery, which has brought many tourists to Bariloche, has deteriorated greatly. Stomach analyses of the "trucha" show that 70% of the food is smaller "trucha." There are three species of deer on the island, all introduced, and the most important of which is the European red deer. Contreras told us that there was a population of one of these per 2 ha on the island, and the effect of their browsing on the vegetation is what one might expect. He said that he and his family had eaten 42 deer since the first of last December!

The biological station was unpretentious but adequate to support several field workers. There were many introduced species of trees near the station including a few sizeable giant sequoia. Contreras explained the presence of the exotics as resulting from the Argentine concept of a national park many years ago being that of a European park. Hence there was no intent to preserve the natural system but only to introduce species that fitted into this concept.

After stopping briefly in Córdoba we were driven by Ismael di Tada to Andalgalá to see the progress of the Desert Scrub research, then returned to Córdoba. En route home we stopped in Cali only to find the university closed and then renewed contacts briefly in Medellín and Bogotá before flying back to the United States.

In October 1971 I flew to Montevideo to take part in the V Congreso Latinoamericano de Zoología, and renewed some contacts en route. In Lima I talked with Bernardo Batievsky in the few minutes that the plane was on the ground there, as he had come to the airport to talk with me. Over a cup of coffee he rapidly clued me in on some of the follow-up on the Lima conference of the previous February on "Environmental Pollution." Progress had been impressive, and he remained "head over heals" in the cause. He was strongly in favor of international standards, citing the example of two Japanese firms that were either moving or hoping to move their operations to Peru because they were so environmentally contaminating that they would not be permitted in Japan. I mentioned the rumors that Brasil would try to sabotage the Stockholm conference. He knew of these, but felt that the efforts would fail. He mentioned good rapport between Peru and Chile with respect to the conference. He mentioned briefly the recent efforts (largely his own, I believe) to cut down the traffic in wild animals from Peru.

I next stopped in Córdoba to check on progress of Mike Mares, my graduate student, who was working on convergent evolution of mammals as a part of the Desert Scrub Program. I also renewed contacts with Jorge Abalos, Ricardo Luti, and Ismael di Tada at the University.

At the Congreso in Montevideo I gave a paper describing the IBP research in the United States under the title "Analisis de Ecosistemas en el Programa Biológico Internacional." Because of a cancellation on one session I was also able to give a paper on my own research on toads.

En route home I stopped in Miami Beach for the II National Biological Congress for which I had organized a symposium session on "Problems of Developing the New World."

The symposium was a real success. We had a good audience and many questions. Ing. Luis Chang Reyes came from Peru in place of

Bernardo Batievsky and presented an excellent paper. One of his strongest points was that the LDC's wanted environmental and health-hazard standards of international scope. He cited export to Peru of cars that because of their design are so polluting of the air that they cannot be sold in their country of manufacture. He also cited the mercury problem in fish caught off the Peruvian coast and presented the proposition that this must have come from great distance, since Peru has no industrial source of mercury pollution.

At the Congress I talked with Dr. Richard Saunier, who was then director of the Peace Corp's Environmental Program in Latin America, with home base in Lima. He said that he had come to the Biological Congress principally because of our symposium on Problems of Developing the New World. We talked principally of two projects that could profitably involve Peace Corps volunteers on either a short-term (six months) basis while awaiting longer-term assignment or on longer-term (two year) assignments. The projects were the Tropical Rainforest at Providencia in Colombia and the Andean, joint HA-ecosystem project in the Andes that Paul Baker and I had been discussing.

Continuing our efforts to develop tropical rainforest research, I arranged for a small group to meet in Medellín in mid-December, 1971, and visit the biological station at Providencia. Tom Odum, who had agreed to try to develop a tropical forest biome project under US/IBP, Bill Milstead from our coordinating office in Austin, Don Hunsaker from U.S. Peace Corps in Bogotá, and I met with scientists and administrators at the University, then went with Fabio Heredia to Providencia. The station was built with U.S./AFOSR funds for joint research on arborviruses by University of Wisconsin and Universidad de Antioquia scientists, but was being abandoned by this project and being turned over to the Universidad de Antioquia to be administered by the Department of Biology. The main advantage of the station was its location in an area having virtually undisturbed forest and on land belonging to a gold mining company and near a hydroelectric plant. The main drawback was a logistical one. To get there we flew to El Bagre on the Caribbean coastal plain, took a one-two hour trip upriver by motorized canoe, then a two-hour ride by pickup truck over a primitive road.

My next trip to South America was again to Medellín in mid-July, 1972, to participate in a pre-planning conference for the Institute of Ecology's workshop on tropical ecology. En route, Fabio Heredia and I met in Bogotá with Dr. Ospina, Director of Colciencias, and an assistant to urge on them the significance of the Providencia station for Colombia and for tropical ecology generally. The result was sufficient financial support from Colciencias to maintain the station after its transfer to the University of Antioquia.

In late February 1973, Jim MacMahon from the Desert Biome and I visited the Origin and Structure of Ecosystems research sites in Argentina and Chile in hope of effecting more cross-communication between the two programs. In November 1974 Fred Wagner from the Desert Biome and I went to Mendoza, Argentina to participate in the V Reunion Nacional para el Estudio de la Zona Afida Latinoamericana. We had hopes of developing grounds for post-IBP cooperation with the Argentines in arid-zone research. Both of us gave talks, and I was asked to participate in two Mesas Redondas, one at the Reunión and one at a luncheon meeting hosted by El Banco de los Andes.

At the latter I emphasized the need for innovative research into use of arid land plants and animals. I used *Larrea* as an example of a possible source of complex chemical compounds. This aroused considerable comment in the middle of my talk, and Vargas from the Mexican arid lands committee talked of Mexican work in this direction. When I got the floor again, I emphasized the need for expanded, cooperative hemispheric research on the structure and function of arid-land ecosystems. I ended by mentioning the NSF cooperative Latin American program.

3

Selling the IBP in Washington

Roger Revelle resigned as chairman of USNC/IBP at the end of 1967. NAS President Frederick Seitz called me and asked if I would be willing to take on the chairmanship. After thinking it over for 24 hours and receiving encouragement from University of Texas President Norman Hackerman, I returned Seitz' call and accepted the chairmanship.

Retiring chairman Revelle and I met the Washington press at a well attended press conference on January 20, 1968. The one thing that I recall vividly from that session was one reporter who kept boring in on the question of "what was the IBP doing about defoliation of Vietnamese forests by the U.S. military."

As I took over the chairmanship there seemed to be several formidable problems facing the US/IBP organization. One was the need to streamline the unwieldly organization into which the U.S. National Committee, its nine subcommittees, plus a task force on Biological Control, a Biometeorology Panel, and an Aerobiology Panel had evolved. The second problem was one of focusing the research planning on a few really integrated research projects, and getting out of the trap of trying to identify individual research projects that were thought pertinent to the overall objectives of the IBP. The third and most difficult problem was one of trying to generate new money in the Federal budget for these large new undertakings.

The first of these problems was relatively easy to solve. A plan for reorganization was presented to NAS early in 1968. When the traditional time (July 1) came for reappointments to NAS/NRC committees, all of the subcommittees and panels except the Biometeorology Panel would be abolished. The National Committee would consist of a small Executive Committee, a 10-man International Coordinating Committee with one person from each of the abolished subcommittees or Aerobiology Pane, a Program Committee (PROCOM) made up of directors of approved or developing IRP's or biomes, and the Biometeorology Panel.

The British journal *Nature* again commented on the US/IBP:

> "With the more recent streamlining of the committee structure, the academy's IBP organization looks more like an executive body and less like a gathering of all the worthies in the field."

The only significant further change in the structure of the U.S. organization for the IBP was made in mid-1971. USNC/IBP became the Executive Committee plus directors of ongoing IRP's and biomes. At the same time it was decided to make a distinction between IRP's (Integrated Research Programs) and CORP's (Coordinated Research Programs). A CORP was defined as, "An association of related research which when taken collectively make a contribution to the field. The emphasis is on coordination of closely related, usually ongoing research." This move ruffled a few feathers, but it seemed necessary to make a distinction between these modest efforts at coordination, eight in all, and the exciting "big biology" that was emerging in the IRP's and biomes.

The second problem, that of focusing the research, was not easy to overcome. In the first year of my chairmanship, we had one full-time staff man in the IBP office at NAS/NRC culling records of research projects, supported by Federal agencies, that could be identified as pertinent to IBP as defined by the numerous U.S. subcommittees and panels. Individual scientists were encouraged to submit descriptions of their research projects to the national committee to be considered for inclusion in the IBP. Since Federal agencies and individuals had been encouraged to approach IBP with their projects, there was a feeling that there was an obligation to list these research projects as a part of the U.S. participation in the IBP. As mentioned earlier, both Part 1 (July 1967) and Part 2 (December 1968) of Report 3 of the USNC/IBP included listings of these IBP-approved projects. The otherwise favorable article on US/IBP in *Nature* of December 2, 1967 referred unfavorably to these listings:

"Apart from these committee-sponsored moves in a grand strategy, the recent report lists over 100 individually proposed schemes relevant to the IBP, many of which are going on anyway and are already financially secure. The number has now risen to a total of about 170. This section of the report is quite a rag-bag, and the items very uneven in value and interest. This section much more resembles the national programmes of other countries such as Britain where the impression is left that the IBP has provided a new system for indexing research projects already in progress."

One of the first tasks I undertook as chairman was to work to shift emphasis to development of the IRP's, and no IBP-related individual projects were approved after February 1968. The listings of such projects in Report 3 hurt the image of the US/IBP, but these were quietly forgotten as the IRP projects took form. I am convinced that the IBP would have been a disaster in the United States, bearing out the pessimistic predictions of Lamont Cole and Tom Park, if we had continued along the lines we were following in late 1967 and early 1968.

At the same time our committee was struggling with the problem of IBP-related projects, we were also faced with the need to determine what IRP's would be accepted as components of the U.S. program. It was difficult to stay within the stated objectives of the IBP in approving the development if IRP's, and in other instances it was difficult to find people willing and able to invest their time in developing IRP's considered to be of high priority but with no certainty of ever being funded.

In the course of 1968 the USNC/IBP considered widely diverse proposals in addition to those that eventually developed into the core of the U.S. effort. These included such diverse subjects as *Chronobiology* (physiological effects of time-zone changes on human subjects), *Biogeography of the Sea, Physiology of Colonizing Species, Crop Production Under Stress,* and *Ecology of Nitrogen.* Approval was given for planning conferences on such additional subjects as *Population Dynamics* (human), *Living Fossils,* and *Thermal Pollution.*

By the end of 1968 the U.S. program was fairly well shaken down into the one for which support in varying degrees of adequacy eventually would be found. In the Human Adaptability component the program included:

(1) *Population Genetics of South American Indians,* directed by J. V. Neel.
(2) *International Study of Circumpolar Peoples,* which was later split into separate Aleut and Eskimo subprograms directed by W. S. Laughlin and F. A. Milan, respectively.
(3) *Biology of Human Populations at High Altitudes,* directed by P. T. Baker.
(4) *Migrant Peoples,* with D. B. Shimkin as Director and latter renamed *Biosocial Adaptation of Migrant and Urban Populations* as a CORE rather than IRP and with E. S. Lee as Director.
(5) *Nutritional Adaptation To the Environment,* which became a CORE, with O. L. Kline as Director.

In the Environmental component the Analysis of Ecosystems IRP had planning under way and directors designated for all of the five biomes, which eventually would be funded individually at considerably higher levels than any IRP. Directors were G. M. Van Dyne (grasslands), S. I. Auerbach (eastern deciduous forest), S. P. Gessel (coniferous forest), D. W. Goodall and F. H. Wagner (desert), and J. Brown (tundra). Efforts continued to be unsuccessful to get support for the Tropical Forest Biome.

Other principal IRP's included my *Convergent and Divergent Evolution,* later renamed *Origin and Structure of Ecosystems,* with three subprograms: *Desert Scrub* directed by Otto Solbrig, *Mediterranean Scrub* directed by H. A. Mooney, and *Island Ecosystem Stability and Evolution* directed by D. Mueller-Dombois. Another ecosystem oriented IRP was

R. C. Dugdale's *Biological Productivity in Upwelling Ecosystems*. The other projects were a rather diverse lot:

(1) *Marine Mammals*, directed by G. C. Ray, considered an IRP but never got support for integrated research.
(2) *Aerobiology*, directed by W. S. Benninghoff and designated a CORE.
(3) *Biological Control of Insect Pests*, directed by C. B. Huffaker, first designated a CORE, but later changed to IRP and its name changed to *Integrated Plant Pest Control*. This later developed into a major program with life beyond the end of IBP.
(4) *Conservation of Plant Genetic Materials*, directed by J. L. Creech and designated a CORE.
(5) *Conservation of Ecosystems*, directed by John Buckley, later by George Sprugel, and eventually by Rezneat Darnell, and designated a CORE.

By the end of 1968 it was evident that the US/IBP was made up almost entirely of large integrated projects with strong emphasis on the ecosystem approach.

The biggest problem that faced us was the one of convincing Federal budget makers of the merits of providing adequate support for the research that was being planned as our participation in the IBP. In some ways this was one of the worst conceivable times to seek funding for a new program, as the Federal budget tended toward austerity. On the other hand, the late 1960's were a period of mounting concern about the environment, which should have been favorable to the obtaining of support for a program with the theme of the IBP.

The Washington climate was good for development of IBP funding, but a long, hard push was ahead. This climate was partly a reaction to a report published in November 1965 entitled, "*Restoring the Quality of Our Environment.*" This in-depth report had been prepared by a prestigious panel of scientists chaired by John W. Tukey of Princeton University and reporting through the President's Science Advisory Committee (PSAC). In his letter of release of the report, President Lyndon Johnson wrote:

"Looking ahead to the increasing challenges of pollution as our population grows and our lives become more urbanized and industrialized, we will need increased basic research in a variety of specific areas, including soil pollution and the effects of air pollutants on man. We must give highest priority of all to increasing the numbers and quality of the scientists and engineers working on problems related to the control and management of pollution.

"I am asking the appropriate Departments and Agencies to consider the recommendations and report to me on the ways in which we can move to cope with the problems cited in the Report. Because of its general interest, I am releasing the report for publication."

The National Science Foundation, like other Federal agencies, responded to the request in the last paragraph of the President's letter. On February 24, 1966, John Cantlon, then doing a stint in Washington as Program Director for Environmental Biology at NSF, wrote to B. H. Ketchum, then President of the Ecological Society of America. He asked that the Ecology Study Committee of the Society provide him with its commentary on the specific recommendations in the report. Among the questions he asked was: "How would IBP rate among the various major themes as a priority label for requesting funds?" Cantlon wanted our commentary by the end of March. I collated and edited the responses from our committee members and on March 28, sent our reply to Cantlon. In reply to his question about IBP my letter read:

"We believe that the IBP is an important part of the theme of environmental quality. The reason we say this is that the basic role of IBP is one of providing an evaluation of the extent of environmental degradation being carried on by mans' activities throughout the world. Moreover, many of the studies in productivity, cycling, conservation that are proposed are in reality baseline studies which will provide us with a description of the environment as it is now and it will provide a benchmark against which future comparisons can be made. Consequently, we would consider IBP as an example of the kind of program which is aimed specifically at fulfilling some of these recommendations and needs posed in the publication by the President's Advisory Committee."

In mid-1966 the Subcommittee on Science, Research, and Development of the House Committee on Science and Astronautics held hearings and in October issued a report entitled, *"Environmental Pollution, A Challenge to Science and Technology."* Chairman of the subcommittee was Congressman Emilio Q. Daddario of Connecticut; chairman of the parent committee was Congressman George P. Miller of California. The report was prepared under the direction of Richard A. Carpenter, a senior specialist in science and technology, Science Policy Research Division of The Library of Congress. We were to see much of all three of these men in coming years, and the success in generating Federal support for the US/IBP is attributable in large part to their interest and support. The first two recommendations in the October report are indicative of the mood of this congressional committee and subcommittee.

"1.   To improve our knowledge of what we are about, scientific activity in ecology and related fields should be immediately expanded to provide-
    (a) Baseline measurements in plant and animal communities and the environment — an ecological survey.
    (b) Continued monitoring of changes in the biosphere.
    (c) Abilities to predict the consequences of man-made changes.
    (d) Early detection of such consequences.
    (e) Knowledge of the environmental determinants of disease.
"2.   Ecological surveys and research should be centralized as to management in one science-based Federal agency. The scientific activity should be performed (whether in Government laboratories or under contract by local universities and research institutes) in geographical regions which correspond generally to natural environmental boundaries."

Roger Revelle had discussed with Congressman Miller the need for official recognition of the IBP by the U.S. Congress. On March 9, 1967, Miller introduced a bill (H. Con. Res. — 273) into the 90th Congress expressing support for the IBP. This bill read as follows:

"Resolved by the House of Representatives (the Senate concurring), that the Congress hereby finds and declares that the International Biological Program, which was established under the auspices of the International Council of Scientific Unions and the International Union of Biological Sciences and which is sponsored in the United States by the National Academy of Sciences and the National Academy of Engineering, will provide a unique and effective means of meeting the urgent need for increased study and research related to biological productivity and human welfare in a changing world environment.

"The Congress commends and endorses the International Biological Program and expresses its support of the United States National Committee and the Interagency Coordinating Committee, which together have the immediate responsibility for planning, coordinating, and carrying out such program in the United States.

"The Congress calls upon the Federal departments and agencies and all persons and organizations, both public and private, to support and cooperate fully with the program and the activities and goals of such Committees."

The same resolution (S. Con. Res. 26) was introduced into the Senate on May 10, by Senator Fred Harris of Oklahoma. In the Senate the bill was routinely referred to the Committee on Foreign Relations, chaired by Senator William Fulbright of Arkansas. This committee had more pressing matters on its mind in 1967 and never got around to a consideration of the bill. The situation was very different in the House, where H.C.R. 273 was referred to Congressman Miller's Committee on Science and Astronautics. Miller gave a free hand to Congressman Daddario, whose Subcommittee on Science, Research, and Development held five hearings on the bill between May 9 and August 9, 1967.

The Daddario subcommittee made every effort to give the IBP full exposure in the five hearings. The first session on May 9 had US/IBP Chairman Roger Revelle, Vice-Chairman Ted Byerly, Harve Carlson of NSF, Fred Smith of the Analysis of Ecosystems Program, and from the Office of Science and Technology, Ivan Bennett, to represent the point of view of the Executive Branch.

Revelle led off with an eloquent and later frequently quoted statement of the rationale of the IBP:

"Our goal should be not to conquer the natural world but to live in harmony with it. To attain this goal we must learn how to control both the external environment and ourselves. Especially, we need to learn how to avoid irreversible change. If we do not, we shall deny to future generations the opportunity to choose the kind of world in which they want to live.

"Greater understanding will make it possible for man to respond to opportunity as well as to react to need. To gain such understanding is the underlying purpose of the International Biological Program."

Bennett spoke favorably of the IBP, and in response to a question from Daddario, "Does your presence here, Dr. Bennett, prove that the administration is in support of this proposal?" Bennett replied, "Yes, indeed."

During the course of these first hearings Fred Smith made a strong case for ecosystem ecology as follows:

"I would like to point out we have been talking about two rather different levels of ecology so far this morning. One concerns physiological ecology, or life history ecology and this is a subject which is well developed in this country and in many others, but is almost totally absent in the underdeveloped countries. This is the kind of ecology we can extend around the world at a relatively low cost and rapidly with an adequate organization.

"The other level that we have been talking about is the ecosystem, and this is not an aspect of a subject the way ecology is. This is a human study itself. It is a whole new scientific level, like a molecule or an organism.

"An ecosystem is a physical entity which can be studied. And I think it is fair to say that in all of the developed countries the functioning of ecosystems has become a matter of national concern in terms of the environment; certainly the semipopular literature of Russia is aware of the problem; they are concerned with water pollution, air pollution, and the destruction of environments. And it has become very evident now that man has become so ubiquitous that he is influencing all natural systems whether he knows it or not. The tragedy is that we do not understand natural systems.

"If we put together what we do know or think we know, we produce models that in fact don't operate.

"So there is an awful lot that we know we don't know. And these are primarily concerned with the checks and balances in natural systems that allow them to persist. We move into these and the systems are changing. We can see them change, but we don't know how far they are going to go on changing; we don't know what we could do to bring them back where they were if we wanted to, or how to control their change in the future.

"It is for these reasons that we need these very intensive analyses of whole ecological systems.

"Now, included in these we will have to have natural systems—'less disturbed ones,' to find out as best we can how the whole process of evolution solved these problems and then also study some that we could upset pretty badly to see what we have been doing to them."

The second hearing was held on June 6, with testimony from S. Dillon Ripley, Secretary of the Smithsonian Institution; by myself, representing the Ecology Study Committee; by Roger Porter, President of AIBS, read by John Olive and by R. L. Zuemer, former director of FASEB. Ripley spoke first and to me his position seemed ambivalent and defeatist, although he professed support for the IBP:

"I must say that I am rather disappointed by the general reaction of biologists in this country to the International Biological Program, and I am also somewhat disappointed by the rather bland approach of this concurrent resolution, which seems to me to reflect the general uncertainty in the minds of both biologists and people concerned with legislation of what the problem really is.

"I think perhaps that to many biologists this general program seems diffuse. They are concerned with immediate research problems in their own laboratories, they are concerned with their immediate careers, they are concerned with grantsmanship, promotions, activities in their universities or laboratories which are entirely personal to them, and so their reaction to the program is much the same as the man on the street's. That it is something that is not quite going to concern them in their time.

"It is particularly difficult to come to grips with broad environmental problems; many of them sound overly intricate, highly theoretical. Some of the ecologists who work on these problems are people who are concerned with such things as mathematical models, and various sorts of stochastic processes in biology which are fringe-like; they are not really central to the main efforts in biology today, and they certainly have no large political voice within the scientific establishment.

"I have been, as I say, somewhat depressed by the lack of response and interest of the biological fraternity in general in this really basic and highly necessary program."

In response to Ripley's comments about the resolution as being bland, Daddario commented:

"You have referred to this resolution as being bland, which causes you concern. Because it is before us to be examined and because we have no commitment to keeping the language of this resolution in its present form, how do we put some salt and pepper to it to make it a little bit more tasty for the whole biological community, and how could we stir up some better appetites for these people who you say are not as involved in these vital areas as they should be?"

Ripley replied:

"I feel that there are two things to which perhaps this resolution should speak. One is that it would urge the biological community in the United States to pay greater attention to the program. There must be ways of making the language a little more hortatory. And the other is there should be some mention of budget.

"I think the heart of the matter really reaches to that. As I tried to say in my testimony, the average, let us say, young biologist on the way up in his career profession lives so much by the rule of grants and grantsmanship, whether the money eventually comes from the Federal Government or State programs or foundations, that he is roughly the way the new doctor was, the new medical man, in the 1940's: worrying about his immediate bread-and-butter problems.

"The nub of the problem here is represented by the fact that there has been no identifiable budgeting for this program so far, and that means that many of the most active minds among young biologists will think that they have to go where the grants are. And there is something to it, I must say, if they are married and looking forward to a successful career.

"There is an inevitable lack of enthusiasm by even the brightest biologists for activities which are not going to be financially or career rewarding. And that is why I called it dirty and formless, because environmental problems are not necessarily those which have an accepted and relatively easy success pattern, as, let's say, molecular biology has at the moment.

"It is easy to be in molecular biology, especially if you are very bright, and to have an assured future. It is not easy to be in environmental biology if you are very bright, and to have an assured future. And it is simply a fiscal problem in a sense as well as a career advancement problem.

"The average young man today wants to go, especially if he is bright, where his career will be probably the most assured and into an area of research where he will presumably have the most success, the most patting on the back eventually, and the most esteem by his colleagues. And this is a pioneer area, and pioneer areas are no more fashionable among highly skilled, trained professional men than they are to people who have to go out and meet any other kind of pioneer trial. So there is money involved in the lack of success of this program so far.

"I think that those two things are perhaps simple explanations of the lack of ability of this program to catch on."

It seemed to me that Ripley was overly cynical in these comments, and when my turn came, I said:

"I think it is not a matter of money to entice people into the program; I think it is a matter of a good program which is emerging—there are several aspects of it that I would like to speak to—which will attract good people into the program. It is attracting good people into the program.

"However, this does not remove the necessity for financial support if we expect to do what we set out to do under IBP."

I also made a strong pitch for cooperation with South American countries:

"Again we are going to set up cooperative programs with the people in South America. Again this will take money. But I think there is a big spinoff from this kind of effort in simply upgrading biological science in Latin America, upgrading our own relations with Latin American biologists. The relations with South American biologists are fine; the communication is terrible in my experience. The people there are enthusiastic about these programs, and I see a chance to perhaps avoid a possible Vietnam, for example, in South America simply by developing this kind of liaison with these people, this kind of cooperative research. I am not talking about expeditions down there to collect specimens, I am talking about working cooperatively with these people. There is a big difference in their mind."

I also stressed the distinction between identifying ongoing research and calling it IBP and the development of the new, large ecosystem studies referred to by Fred Smith:

"Now there are aspects of IBP that have simply consisted of identifying on-going research and the Government agencies identifying individual projects by various people and saying well, this fits what we are talking about in IBP so this is part of our IBP program. Well, I think this is pure bookkeeping.

"What I am thinking about is these programmatic things that could not be done without a big effort, and these are the ones that will cost money."

Congressman George E. Brown of California, one of the more articulate members of the subcommittee, picked up my train of thought:

"Looking at the statement which Dr. Blair has submitted—on page 4, the last paragraph—I think we have a key which, in my opinion at least, can establish the base for justification of the funding when he says that: 'The IBP effort of the United States can serve as an effective umbrella for many of the proposed activities related to environmental quality and baseline environmental studies with which the Congress is currently concerned. Virtually all of the bills before the Congress that are directed toward problems of environment quality are directed toward problems that are encompassed by the U.S. plan for participation in the IBP.'

"Isn't what we are talking about, then, the R & D component of this tremendous program of environmental control which is involved in air pollution, water pollution, ground pollution, and many other things which has become catastrophic in its nature but about which we have lacked any kind of consistent understanding of the basic factors?

"If the IBP will provide the R & D component, it could, in itself, save many times the amount of money that would be required to justify the action which is going into these things."

The subsequent statements by John Olive on behalf of Roger Porter and the prepared statement by R. L. Zwemer were supportive of the IBP.

The third hearing was convened on July 12 again with Ivan Bennett, plus Stanley Cain, then Assistant Secretary, Department of the Interior and Vice-Chairman USNC/IBP, David Keck of NSF, and B. H. Ketchum of Woods Hole Oceanographic Institution and USNC/IBP.

Cain testified:

"I find in Interior a deep desire to carry on more fundamental interdisciplinary ecological types of investigations as related to production of ecosystems, as related to their structure and functioning, as related to their geographic distribution and ultimately as economic species that are parts of this system, but we need this fundamental knowledge and the mission agencies recognize this.

"There are two points to emphasize.

"One is that the scientific community has gotten tremendously excited about this as something worth doing. These are already all busy men. They have been doing science. That is their life.

"Now, they want to do this sort of thing. Some of them have been doing this and see an opportunity to do more. Others have perhaps been in other activities and want to get on this because it is exciting to them and looks very much worth doing."

Cain estimated the cost of U.S. participation in the IBP at $136 million and further stated in his prepared statement:

"It seems equally clear to me that not all of the funds being sought by American scientists to support IBP research can be squeezed out of existing appropriations for the agencies. That kind of latitude just does not exist. Furthermore, as important as we feel that this program is for national goals, it should not be entirely funded at the expense of existing worthy programs."

Both Bennett and Ketchum spoke in support of the IBP. Dave Keck spoke to the needs for new funding:

"We have looked in the National Science Foundation at the kinds of projects that we are supporting that are 'IBP-like' and in fiscal 1967 we supported about $4 million of these projects. Most of those scientists will now shift emphasis and get full swing into IBP projects.

"Those same people are going to need money under the IBP banner, but also the IBP program as we now see it crystallizing is definitely going to need new money beyond and above anything we can see in our National Science Foundation present or future budgets based upon the growth rate we have been experiencing. We feel in the Biological and Medical Sciences Division in the National Science Foundation that this demand, to fund IBP projects which are first rate and that should be funded through our Division, may be something of the order of $20 to $25 million for brand new money, using your term. Mr. Brown, that we do not have in hand now.

"Somehow, we need to get this into the system if the IBP is going to flourish and move ahead at the rate it should."

Congressman Brown responded with a statement that must have produced some qualms among physical scientists:

"You are aware that the committee in general has been quite favorably disposed toward the type of program we are talking about. I know I am personally. We are in a period in which programs which require the identification of even small amounts of new money are going to have tough sledding and I think we would look as a possibility at the prospect for perhaps rechanneling some funds which, for example, are going into the physical sciences at the present time.

"Maybe this is heresy, but I think some of these statements which have been presented this morning indicate that there is an imbalance as between the two areas and instead of looking at each new prospect as something on which we have to load new money, is there some possibility that some of the funds, some of the emphasis, can be changed as between the physical and the biological sciences? Some of these quite large amounts which are now going into a little more esoteric phases of physics and chemistry and engineering could be diverted into the biological sciences."

The fourth session heard testimony from David Gates, then Director Missouri Botanical Garden. Gates approached the IBP matter in context of priorities in science:

"We must arrange some priorities in science, or at least if not priorities let's work on some of the urgent matters for which passage of time means lost opportunity. If many ecosystems are not studied vigorously and soon, then it will be too late — they will be gone. I want to make it emphatically clear that I am not saying that we should not study the nucleus, the stars, and the galaxies — we must study those things, also. We also must and should build new nuclear accelerators. But the study of natural habitats is urgent and critical. No matter what else we do — no matter how many nuclear powerplants we build, how many trips to the Moon or to Mars or Venus we take, or how many rivers we dam and divert — we must still live with and within the natural history ecosystem of the surface of the Earth. We are in desperate need to understand this point now.

"Why do we do those things first which are less urgent than others? When will we realize that soon it will be too late to study relatively undisturbed plant and animal communities? Is it more urgent to study the galaxies, the stars, the planetary systems, which will be here in a thousand years hence, than it is to study the biota of the terrestrial habitats? The terrestrial ecosystems are the most susceptible of all to destruction.

"The aquatic systems — the oceans, the lakes, the rivers — are next. They have a little longer time constant, particularly the oceans, the change is a little more sluggish, a little slower, but there is damage, there is change while the nucleus, the stars, the planets, the solar system and the galaxies, they will be here in-definitely. Yet, what we are doing is to study these things with considerable emphasis, and to delay or to dabble with studying the terrestrial ecology. I do not wish to be misunderstood. I am not saying, let us not study the stars and the galaxy and the nucleus — we must. But if there has to be a choice, if there is a matter of urgency, the terrestrial system, the surface of the earth in which we are living, the thing that surrounds us on all sides, that is a matter of urgency. This is where the big changes are. This is where the impact of man is occurring."

In response to a question from Daddario, concerning the need for theoretical ecologists, Gates replied:

"I envision quite a new thing that has not existed, and that is people trained in ecology as it now exists with strong fundamental work in biology, but in addition good training in physics, and good training in mathematics and the use of computers. By theoretical ecology, I mean those people who can take the data that is accumulated in the field from observation and bring it onto their desk and organize it and derive from it causes and effects and out of this theory. Hypotheses and theories really pull all the threads together into a coherent fabric which will give ecology a real body.

"You see, the difficulty at the present time — this subject is so complex and so diverse if you look in the literature, there are vast amounts of data, of wind, of radiation, of moisture, of water, of plants and animals and all the characteristics, and that is it. It is descriptive. We need to pull this all together into coherent models that show the events, the cause and effect, the way things go, the laws that exist in ecology, whether it is population dynamics, whether it is the food chain, whether it is an interaction of the atmosphere with the plants and the animals; I happen to be involved in this quite deeply in my own work and that is why I feel so strongly. This, in fact, is the reason I gave the introduction that I did, that I was trained in physics, experienced in geophysics and came back into ecology."

The fifth and final 1967 hearing heard Sid Galler, Assistant Secretary, Smithsonian Institution and preeminent Washington biopolitician, and Carleton Ray from the US/IBP. Galler's testimony was somewhat ambivalent but generally supportive of the IBP. His most important point was:

"It is my carefully considered opinion that the U.S. International Biological Program will flourish only if an adequate fund is established for its support, a fund that is basically independent of the annual appropriations for the support of the mission-oriented research programs of the R & D agencies."

Ray testified in support of the IBP and argued for an IBP appropriation:

"Now, specifically, why a special appropriation for the IBP? For one thing, it is perfectly possible to have a program approved by IBP, yet turned down by one or many of the funding agencies in the Federal or private systems. Many or most of the IBP's aims are not appropriate for these agencies which tend to be much more jurisdictional and provincial than the IBP calls for. For another, without being specific, it is not good in practice to force such large enterprises as IBP to go begging to agencies which too often have their own interests to defend and which do not like doing the bookkeeping and financing for other agencies.

"No, I do not suggest or even foresee the creation of another empire. But I do see the need to fund world problems differently than more local ones, and I do see the absolute necessity for putting some toothy funds behind House Concurrent Resolution 273, which endorsed the IBP. I'm afraid we are going to look pretty silly if we don't. The conflicts of interest will be too obvious and the program will not, I would venture to say, achieve wide support from the biologists.

"The funds most needed are those for ecological and other field-oriented research, for training of personnel, and for data-sorting centers. There should be a real effort made to keep the overhead minimal and Parkinsonism at bay. As things stand today, however, the IBP has no real teeth, can endorse, but not fund, and cannot possibly do the job it has set out. It is a bit like trying to build a good pie with 100 bakers directed by a committee. The IBP's pie needs the vintage fruits of research from an oven stoked with the wherewithal of a congressional appropriation."

As a result of this extensive series of hearings, Daddario and his largely sympathetic subcommittee decided that the best way to advance the IBP to adequate support in the Federal budget would be to prepare a special committee study of the IBP. This was published in March 1968, under the title: "The International Biological Program: Its Meaning and Needs." Richard Carpenter, Freeman Quimby, and Marcia Carlin of the Science Policy Research Division, Library of Congress prepared this document, which, in my view, was largely influential in ensuring Federal financial support for US/IBP.

This document made a good case for the IBP. It also included these recommendations:

"A firm commitment by the U.S. Government to support the IBP to the maximum feasible extent and for the full 5-year term, longer if necessary. This should be done both through Executive expression and legislative resolution.

"(2) A determination, agreed upon with the Bureau of the Budget, as to the best method of future financing. This may be handled through a single annual line item in the budget of one agency, such as the National Science Foundation, or through the budgets of several agencies. Full exploration of the possibilities should be made. Whatever method is reached, however, should be one that can be depended upon.

"(3) Support by the IBP, the Government, and the academic community for the development of research leading to a theoretical ecology. At the present time almost no such discipline exists; yet, according to testimony, a real understanding of planetary ecosystems is dependent on the development of such theory. The situation appears comparable to that which existed in the 1920's and 1930's when a need for the development of theoretical physics was prerequisite to the emergence of atomic energy.

"(4) That all appropriate Federal agencies support and participate in the IBP by providing use and assistance in the use of their facilities to the United States and abroad. Federal agencies should also continue to review, reorient, and support those in-house research projects which are consistent with the aims and objectives of the IBP, as well as appoint their most responsible and knowledgeable biologists to the various IBP organizational committees or units, as required."

This report quite frankly identified the main problems facing the US/IBP:

"(1) The loose structure and organization of the IBP administrative machinery. — This seemed to involve the management of the entire effort and, very possibly, may have stemmed from an excessive concern over too much centralization. Part IV of this report describes the organization in some detail. Suffice to say here that a more tightly knit and cohesive managerial arrangement, including a centrally located executive group with personnel and staffing not committed to double or triple duty, seems essential.

"(2) An inadequate and unrealistic mode of funding. — The foregoing comments on the financing of the IBP, the subcommittee believes, tell their own story. It is difficult to see how the current makeshift system of providing funds for the U.S. program can result in any reasonably sustained effort. The budget squeeze, which is common throughout Government at the present time, makes this problem especially acute. But it will have to be resolved in some way other than that presently in operation if any progress is to be made.

"(3) The shortage of trained manpower. — No group, obviously, is more aware of this obstacle than the U.S. National Committee for the IBP and those who are endeavoring to make the program viable. While biology is a rapidly expanding field, and in some ways is the 'glamour' discipline of the day, its main thrust has not been toward ecology and its overall research support still lags in relation to its significance and potential—at least in terms of the amount of dollars committed to it. The number of well-trained ecologists is woefully small, and the discipline itself needs much study and development. The American IBP effort is directed partly toward rectifying this situation which, in the final analysis, is a limiting factor and which cannot be cured without desire, money, and time. Since time is of special importance here, delays in ecological training cannot very well be tolerated.

"(4) Lack of public understanding and of firm endorsement by Congress and the Executive.—It goes without saying that this is the most restrictive element which faces U.S. participation in the IBP. To date, the significance of the IBP does not appear to have been brought home to the country in general or to the Federal Government in particular. In view of the urgency involved, concentration on the latter may be warranted since it is more readily reached. Such concern as has been evidenced thus far by the Government—and that concern is relatively mild—seems to have been fostered more by the conditioned response of Government to the prestige of the scientific community than to an understanding of the problem itself. This situation must change—or the IBP is not likely to get off the ground."

While all of this was going on, the Ecological Society of America was getting a part of the action. At the annual meeting in College Station, Texas, in August 1967, I was named Chairman of the Public Affairs Committee with Stanley Auerbach, Robert F. Inger, and David Gates as members.

We held two meetings in Saint Louis on November 1967 and January 1968, to plan our strategy and then tackled the Washington scene.

Subsequent to its activation in August 1967, our Public Affairs Committee of the Ecological Society of America maintained an active stance in Washington through the following year, a year critical for the promotion of the IBP in the United States.

We held our first meeting in Washington on October 25, with conferences planned with various Washington administrators. One meeting was with Mr. Jansen of the Federal Water Pollution Control Administration, who had been given the job of preparing a report on the condition of the nation's estuaries. We indicated our interest in this effort and offered to provide ecological advice both at the level of case history studies and at the level of ecological input into the final document.

Another meeting was with Donald King, a biologist from the Department of Agriculture assigned to the Office of Science and Technology. His main duty at that time was to see through to completion a report of the Federal Council on Science and Technology, originated by Jared Davis of the AEC when he was assigned to that office and since worked over by representatives of the various agencies. Our committee expressed our concern about the watering-down and softening-up of the report that was rumored to be happening as a result of the working-over it was being given by the various governmental agencies.

The report was later issued (March 1968) under the signature of Donald F. Hornig, Director, OST and Charles L. Schultz, Director, Bureau of the Budget. The title of this document was "Advancing Scientific Understanding of Natural Communities." The main recommendations were: (1) designation of additional protected natural areas for scientific purposes, (2) expansion of research, including cooperative studies of a limited number of large ecological systems, (3) expansion of monitoring and survey programs to measure the impact of environmental changes, (4) improvement of ecologically-oriented training in education programs. The final document was much stronger than we had anticipated, but whether or not our Public Affairs Committee could claim any credit for influencing the people in OST was a moot question. We did know that the people we talked to there were responsive to criticisms expressed during our conversations.

During this visit the Committee also visited with W. G. Allen, Staff Assistant to Senator Gore of Tennessee, and talked about the objectives of Senate Resolution 26, the IBP resolution. Finally, we visited with Freeman Quimby and Richard Carpenter in the Legislative Research Branch of the Library of Congress.

Carpenter, Quimby, and Marcia Carlin proved enormously helpful in our efforts to establish contacts on the Washington scene. On my frequent trips to Washington during this period I tried to have lunch with all or part of this trio, where, over hot pastrami and beer at a little restaurant near the Library, we assessed progress and setbacks in advancing the IBP and environmental science generally at the Federal level. Quimby was generally the realist and the pessimist in the group. During our first meeting in Washington, it was decided to meet for a day in St. Louis to prepare a letter to Donald Hornig (President's Science Advisor), stating some of the views of our committee relative to national problems that had an ecological context.

This meeting was held for a day on November 13 in St. Louis where we were able to work in the quiet atmosphere of the Arboretum of the Missouri Botanical Garden, hosted by David Gates. The letter was mailed on November 15 to Hornig, and, as a result of the letter, a meeting was set for December 12 with him in Washington. The letter read:

"The Public Affairs Committee of the Ecological Society of America has been considering the role of ecology and ecologists in environmental problems.

"May we impose upon you to give you the substance of our thinking so far?

"Even if, for a moment, we put aside social and political desires and activities, man lives in an extremely complex world, one in which so many natural forces are interacting that it is sometimes difficult to distinguish between cause and effect. When man's population was small and his technology primitive, his economic activities had only local and usually temporary effects on his environment. Now, however, his population is so large and his technology so powerful that it is quite possible for some of these adverse effects to be global and irreversible. As a result, the most pressing long-range problem faced by man is how to step into this complicated network of interacting forces and extract from it the resources needed for decent human existence without destroying the system that sustains him.

"This problem is one of economics. It is also a major problem in ecology, which well could be called biological economics. Clearly a satisfactory solution must involve political and social considerations. Nevertheless, any proposed solution that is not consonant with the inherent ecological limitations of the world system will be only a stop-gap measure and doomed to failure. In view of the seriousness of the basic problem, society must bring to bear on it as much skill and understanding as possible.

"We believe that ecologists represent a critical part of the talent required and that ecology is a biological discipline provides the theoretical framework within which a solution must lie. Ecologists as individual citizens are more concerned than ever about deterioration of the environment. They are increasingly concerned about the part they, as professional biologists, may play in seeking solutions to these problems.

"Ecologists recognize the need to show the relevance of ecology to these environmental problems. In turn, we hope that public officials will agree that ecologists and their science should participate in seeking solutions.

"Among the many ecological problems facing man, the following are typical but by no means all-inclusive: (1) effects of deliberate and inadvertent weather modification on vegetation and animals, (2) effects of addition of one billion people to the world's population, (3) ecological effects of extinction of many species of plants and animals, (4) pollution of air, water, and land. Solutions to these problems require application of ecological principles. In many cases our sophistication is adequate to the task. In the most complex situations, however, our understanding is not yet sufficiently advanced to enable society to make wise decisions. The most serious weakness is lack of knowledge of the structure, operation, and evolution of ecosystems—those intricate combinations of biotic communities and their physical environments which are unitary reacting systems.

"As soon as we consider ecosystems, basic ecology converges on practical problems. We may illustrate this convergence by posing a general question: What kinds of biological systems are appropriate for man to harvest and how intensively may they be cropped without destroying them? Put in concrete terms, is it better to fish for red snapper or herring? How much fishing pressure can these two very different fisheries stand? Is it better to plant single or mixed cultures of grain? How should the strategies differ in different climates and topographies? Answers to these questions go to the heart of the dynamics of ecosystems.

"What are the best ways of using ecologists and how do we go about actually putting them to work.

"Since the Federal Government is one of the most potent forces affecting our environment, either directly through its action agencies or indirectly through support of programs, a review of its total effort is desirable. We are concerned with a review only in the context of environmental quality. For example, consider the impact of recommendations of the Department of Agriculture regarding the widespread use of fertilizers and antibiotics. What are the true costs of using these materials, as opposed to calculations limited to the product of the farms? What are the overall effects of the Corps of Engineers' program of flood control? What are the real cost-benefit ratios of that program when all effects are considered? What are the side effects of the Bureau of Reclamation's weather modification activities in western mountains? We do not pick these examples in order to single out particular agencies; they simply illustrate federal programs that affect vast areas. A general review of this sort should be very useful to both the legislative and executive branches of government.

"The Committee on Environmental Quality of the Federal Council on Science and Technology has been considering federal programs in this light. It is apparent to an observer of this committee's deliberations that it will be extremely difficult for the committee to render an objective evaluation. The members of the Committee are representatives of the federal agencies responsible for the programs being reviewed. Agency loyalties dominate the deliberations. We say this not in criticism of the hard-working men involved. These men are personally committed to the programs they are reviewing. They have careers at stake. Each participant recognizes these forces and is deferential to the sensitivities of the other agency representatives. Under these circumstances, a man would have to be superhuman to be objective.

"The same forces are at work in other inter-agency committees established to review such things as the use of pesticides and antibiotics. It is the operation of these unavoidable sources of bias that prompt us to recommend a review by persons outside of government.

"But more than a single, external review is needed. Agency programs continue. Periodic evaluation, therefore, becomes a necessity. The PSAC report on "Restoring the Quality of our Environment" was an excellent basic study. There is a real possibility that that effort will be largely dissipated because of no follow-up, objective review. To meet this need for periodic, objective, external review of the impact of federal programs on environmental quality we recommend that the FCST establish a review committee consisting of ecologists and environmental scientists from outside of government.

"The second use to which ecologists should be put is as research workers developing our understanding of ecological phenomena. It may be argued that this is what ecologists have always done. That is true, and we should continue to support ecological research since it has resulted in considerable understanding so far. But a new ingredient is required at this point in our history. Special funds are required to support certain areas of ecological research, which is the equivalent of suggesting that more ecologists be induced to work on problems with rather immediate practical implications. Without this understanding wise political and economic decisions are not possible. To increase our understanding the following needs must be met: Increased support for the training of ecologists and technicians; increased support for basic research in the universities; preservation of natural areas for use in research and environmental baseline studies; immediate implementation of the International Biological Program as an important umbrella for environmental research and training relating to the quality of our environment.

"To implement these we suggest stronger support for academic programs, establishment of a National Institute of Ecology administratively independent of the federal government, and strong financial support for the International Biological Program.

"We hope that we may meet with you to discuss the contents of this letter. Will it be possible for us to see you shortly after December 10?"

Also, on November 15, in response to a request from Congressman Daddario, our committee mailed to him a critique of the hearings which his Subcommittee had held on the IBP.

On December 12, our Committee met with Donald Hornig, and with Ivan Bennett, his Deputy Director. We had an extended dialogue of about an hour during which four major points were discussed. First, we discussed the professional background of the people who make up the President's Science Advisory Committee (PSAC) and we made the point, which was well taken, that 12 of the 16 appointed members are either physicists or in the field of engineering, rather closely related to physics. We urged that consideration be given to the appointment of an ecologist to PSAC. Second, we discussed the need for a review committee external to the Federal Government on protection of our ecosystems and on the maintenance of environmental quality. Third, we discussed many problems of environmental quality and of the relation of IBP to the national interests in environmental quality. We discussed the problems of financial support for the IBP and discussed also the present status of the U.S. program. Fourth, we discussed a National Institute of Ecology.

As a follow-up on the December 12 meeting with Hornig and Ivan Bennett, our committee sent a letter to Hornig on January 29, 1968. It read:

"I would like to express the appreciation of our committee for the very useful dialogue we were able to have with you and Dr. Bennett on December 12. Since that time, we have been giving careful consideration to the major points that were covered in our discussions. We would like to take this opportunity to review some of these points with you.

"Because of the increasing concern for environmental quality and concern for acquisition of ability to predict the effects of man's activities on the earth's ecosystems, we feel the need for a person with an understanding of basic ecological principles on PSAC.

"One such person would be Dr. John F. Reed, President, Fort Lewis College, Durango, Colorado. Dr. Reed is a distinguished plant ecologist with extensive experience in many parts of the United States and Africa. He is a former acting President of the University of New Hampshire. He is a former President of the Ecological Society of America and a former Chairman of the Ecology Study Committee of that society. He has been a member of the Advisory Committee for Environmental

Biology in the National Science Foundation and a member of other committees for agencies in Washington. He has a broad outlook respecting ecological principle and human affairs that we feel would qualify him very highly.

"Dr. John E. Cantlon, Professor of Botany, Michigan State University, is another highly qualified person. Dr. Cantlon served a very successful term as Program Director for Environmental Biology in the National Science Foundation, and in that position demonstrated a very broad perspective in regard to the relations of basic ecology and human affairs. Dr. Cantlon is President-elect of the Ecological Society of America.

"Dr. Donald S. Farner is Professor of Zoology and Chairman of the Department of Zoology at the University of Washington. He is a former Dean of the Graduate School at Washington State University. Dr. Farner is an eminent physiological ecologist with international interests and with involvement in international programs.

"Dr. Stanley A. Cain will be returning to his Professorship of Natural Resources at the University of Michigan after having served as Assistant Secretary of Interior for Parks and Wildlife. Dr. Cain is also a distinguished ecologist with broad experience particularly in North and South America and with broad knowledge of applied aspects of ecology as well as with the various aspects of governmental involvement in problems that have ecological relations. He is a former President of the Ecological Society of America.

"We also recommend the immediate appointment of an external review committee, the appointments to be made by the director of OST. This committee should comprise persons cognizant of basic ecological principles and who have demonstrated their concern for the application of ecological principles to human affairs and with the interactions of our society with the environment which we occupy. We would suggest that this committee consist of nine to twelve members and that it include people from the academic and business communities. We recommend that there be no federal representation on the committee, which would be consistent with the pattern of organization of the President's Science Advisory Committee. The charge to this committee would be to advise the President's Science Advisory Committee, by establishment of *ad hoc* committees of experts to provide advice in depth on specific topics as demanded. You mentioned in our discussions "bite-size topics" for discussion. Such a committee could set priorities for the establishment of the *ad hoc* committees to examine these bite-size chunks. We recognize that such a committee may come into being through the agency of

any one of several bills now in the Congress. However, we feel that the need for such a committee is so urgent that one should be established in OST until one may be implemented through one of the bills now in Congress.

"We would like to set out below some of the bite-size chunks to which we feel this committee might address itself and for which purpose it might set up *ad hoc* committees.

"(1) *C/B ratios in federal programs that affect environmental quality*. We feel that it is very important that the basis for calculating cost-benefit ratios in federal programs such as those of the Corps of Engineers and other federal agencies be reexamined so that true costs, as opposed to direct, short-run estimates, and environmental values besides those immediately involved in a project or program enter into the calculations. We wish to avoid discovering in the future that 'the much of the thrift of yesterday turns out today to prodigality because the price tag did not include all of the social and economic costs' (Senator Henry W. Jackson, *Bioscience*, Dec. 1967). Present practices frequently fail in this regard.

"(2) *A review of how much ecological thinking enters into governmental activities in developing countries*. Many of these countries are tropical ones in which the ecological problems are quite different from those in the temperate zones where much of our agricultural and other technology has developed. It seems quite important to review the application of ecological principles in activities being carried on by organizations such as our AID.

"(3) *Ecological thinking in agricultural practices*. Agricultural practices are ones that have tended to grow through trial and error. The time has come when it is necessary to think more precisely in terms of basic ecological principles in future agricultural practices if environmental quality is to be maintained.

"(4) *Significance of the tropical rainforest*. It has been stated in a recent symposium that the industrialized nations of the middle latitudes may find it necessary to subsidize the maintenance of the tropical rainforests of the earth as a means of providing oxygen for our ecosphere. A panel of experts should evaluate that forecast and try to predict whether or not it is realistic to take steps in this direction. This seems of highest priority if the forecast is justified.

"(5) *Ecological review of practices in the management of fish and wildlife*.

"(6) *The ecology of National Parks relative to public use, the modification of the environment by public use, and the responsibility of the National Park Service for maintaining our National Parks as ecological resources*.

"(7) *The ecological basis for pest control problems.* While there have been various panels dealing with problems created by pest control programs, there has never been one to examine the basic ecological justification in such programs. For example, a committee of NAS-NRC now active in assessing the use of persistent pesticides has one ecologist and a large number of people who might be considered to have vested interests in the nature of their report.

"(8) *Ecological implications of future large scale irrigation of the world's deserts for food production.*

"We hope that these suggestions will be useful to you."

On February 7 and 8, 1968, our Committee met again in Washington in order to make an input into the thinking of various people in the Legislative Branch who were primarily interested in environmental matters. Before going to Capitol Hill, however, we met with Jared Davis of AEC, who possessed considerable experience as an assignee to OST. The talks with Jerry were very useful in giving us an insight into how matters are handled in Washington. We talked also with the staff assistants of several senators, including Senators Jackson, Nelson, Harris, Magnuson, and Muskie.

In all of these talks, we mentioned the IBP and its possible contribution to the solution of the nation's and the world's environmental problems.

The next move by Congressional supporters of the IBP was introduction of H. J. R. 1240 by Congressman George Miller. This was similar in many ways to the now defunct H.C.R. 273, but differed significantly in authorizing $5 million for the IBP in the fiscal year 1969 budget and providing general authorization for the following four years. The resolution was introduced on April 24, 1968, and referred to Miller's Committee on Science and Astronautics. No time was lost in calling hearings before Daddario's Subcommittee on Science, Research, and Development on May 1-2.

In opening the hearings, Daddario stated:

"I should like to emphasize we are not now considering the merits or the desirability of the IBP itself. That determination has already been made. So far as the subcommittee is concerned, the IBP is a promising attempt to deal with an urgent problem; namely, the understanding of our planetary ecology. Without such understanding, we are in grave danger of an accelerating deterioration of the ecosystems on which all life depends.

"What we are concerned with here is the method of implementing the U.S. effort. How can we give the program the support it must have to be effective?"

I led off the testimony with a prepared statement for the record, then spoke to the four problems identified in the committee's report of March on "The International Biological Program: Its Meaning and Needs." My main points were:

"First, I would like to speak to the four problems which were identified by the subcommittee. The first refers to the 'loose structure and organization of the IBP administrative machinery.'

"At the time the report was written, this was a valid criticism. I believe our organization has reacted to this and to other similar criticisms to the extent we now have completely restructured our organization into what appears to be an efficient operational body, and I would like to speak to this in a bit of detail since I think it is a very pertinent matter.

"The recommendation has been made to the National Academy of Sciences-National Research Council and has been approved in the Division of Biology and Agriculture. We anticipate that it will not be treated otherwise and will be approved farther up the line that the following organization operate our USA/IBP effort.

"There will be a five- to six-man Executive Committee which will be the prime executive committee for the US/IBP. This Executive Committee will be answerable to the Division of Biology and Agriculture and to the Academy. Actually, we now have a five-man committee consisting of myself, Dr. Cain and Dr. Byerly, who are the co-vice-chairmen, Dr. Fred Smith who is heading up one of our main thrusts in the US/IBP effort, and Dr. Fred Sargent who represents the human adaptability side of our U.S. effort, which is the other main thrust. This will be the operational committee that will actually be the executive group.

"Now, in essential replacement of the old, very large and very sprawling U.S. National Committee, which actually has done the job of planning, which is now essentially completed, we have recommended a nine- to ten-man committee that we are calling the International Coordination Committee. In other words, this committee will coordinate with the Special International Committee for the IBP. This would have nine to 10 people representing the subject areas of the international committees — the seven that the international organization recognizes and the two we have added in our own program (the Environmental Physiology and the Systematic and Biogeography committees).

"This committee would be charged mainly with developing international coordination of our efforts that emerge from our own program.

"The people we are thinking about as members of the committee are people who already have international contacts, who have international recognition in their fields, and who would be effective and interested in increasing the international coordination.

"Another committee — and both of these committees, this one and the one I will mention next are, of course, answerable to the Executive Committee and on up through the channels — will be made up of the program directors for the integrated programs. These are the people who are active and who are heading up our major efforts in the US/IBP. They would be mainly operating to coordinate the entire program efforts in the U.S. and to interact with the International Coordination Committee to feed their efforts into the international effort, too.

"The second question raised was one of 'an inadequate and unrealistic mode of funding.' Of course, that is what this meeting is all about, I believe. My own position is that the proposal that IBP be funded mainly through a line item in the NSF budget is the cleanest, simplest, most realistic way to handle this funding.

"The shortage of trained manpower — our experience in the development of our integrated programs (the large programmatic programs that are really the core and heart of our IBP participation in this country) has been otherwise than to show that we are too short of manpower to get these programs started.

"Now, the second question I would like to speak to is the one of why we do need to get started now, and I have about five points in all that I will make very quickly. I think they are all points that should be taken into consideration.

"The planning of the United States for participation in IBP has now reached the stage where action is necessary soon, or we are going to lose momentum. The momentum is rather considerable right now. People are interested. They are making plans for participation, and I think we will definitely lose momentum if we do not get more of the research that has been planned now into the action stage. Many of the scientists who have made plans to participate are going to have to turn to other things, the realities of scientific existence being what they are, if they don't get support.

"Just having recently returned from the General Assembly of IBP in Varna, Bulgaria, where 37 countries were represented, I feel strongly that the United States is going to lose its world leadership if we don't get our program developed rather rapidly. It was very obvious at Varna that the whole world looks to the United States in this field for leadership. A part of this has been the admiration they have shown for this program of ours because it does go beyond just reclassifying a lot of ongoing research and calling it IBP. It is definitely a new kind of biology which is admired by these other countries, and I think we are going to lose our leadership unless we show that after having planned this we can go ahead and 'get the show on the road.'

"One of the things that I think is of interest here that came up at the meeting in Varna, Bulgaria was the formation of a small three-man international committee, of which I am a member, to look into the possibility and the problems of establishing a world network for baseline monitoring of our environment in connection with the IBP stations. This came originally from the Swedish delegation. It found very enthusiastic reception by the Russian delegation and by the United Kingdom delegation, and we supported it strongly because it fits our whole thinking in this country. We recognize the need for providing the kind of baseline information that has been requested many times recently by the Congress with respect to what is happening to the quality of our environment.

"If we drag our feet, however, we risk our chance of being influential in developing this kind of operation on an international scale. Then I would also say without being either an alarmist or dramatic that I feel there is a real environmental crisis. The things I just spoke about in Latin America as well as the things that are happening here put pressure on us, I think, to begin to understand the functioning of our ecosystems as quickly as we can. I don't want to be dramatic, but I really think we are reaching the stage where time is running out on us, and I think this committee also has expressed these sentiments very strongly.

"I would also say that one of the five official years of IBP will have passed by the beginning of fiscal year 1969. A year has gone of the 5-year period in which there is a certainty of an international body to overview this international effort."

Ivan Bennett followed with a statement generally supportive of the IBP, for example:

"I wish to emphasize, however, that OST's general views of the potential of the IBP are unchanged and the strength of OST's support for U.S. participation in the IBP has not waned during the past year."

He then stunned the supporters of IBP by saying:

"In summary, Mr. Chairman, for the reasons that I have outlined, OST is *opposed* to the resolution under consideration by the subcommittee."

It seemed obvious that he had gotten orders from higher up in the Johnson administration to oppose a new program that would require additional Federal expenditures.

Bennett did say, however, that:

"We will continue to follow closely the progress of the IBP. While we cannot give assurances that it will be funded at any particular level during fiscal year 1970, we have taken steps to assure that it will be fully considered. OST has worked with the Bureau of the Budget and NSF in developing the outline for a special study which will lead to a program memorandum on the IBP for submission by NSF to BOB in connection with NSF's fiscal year 1970 budget request. NSF will have responsibility for the study but undoubtedly will be assisted by USNC/IBP and by other agencies, probably through ICC. The special study and program memorandum will deal with the goal and objectives of the program, the progress of USNC/IBP in developing program proposals, the organization and management arrangements for carrying out the proposed studies, the scientific and other benefits that can be expected at alternative levels of funding for the program, the ability of the 10 or more agencies that have an interest in parts of the IBP to participate in and support the program, and the necessity for any special funding that NSF or other agencies feel is warranted for the program."

The next witness was Leland J. Haworth, Director of the National Science Foundation. Haworth spoke in support of the IBP:

"Without hesitation, we endorse the International Biological Program and the desire behind House Joint Resolution 1240 to give adequate financial support to it. With respect to the express provisions of the resolution regarding funding, however, we defer to the positions being taken by the Bureau and the Office of Science and Technology, with respect to additional specifically identified funds in fiscal year 1969.

Our own budget for fiscal year 1969 as submitted to Congress contains a modest item for IBP. This is the only such line item in the budget of any agency approved for submission by the Bureau of the Budget. However, as Dr. Bennett has said, the Bureau has recently requested us, with a target date in August of this year, to prepare a program memorandum looking toward appropriate

funding levels for IBP in fiscal year 1970. According to the guidelines suggested by the bureau, problems of interagency coordination and funding must be identified and at least tentative
resolutions proposed; there is no intention that the Foundation
should receive all of any new money recommended for IBP."

As in the case of Bennett, Haworth was in no position to oppose the
Administration opposition to the new funds in the amount of $5 million for
fiscal year 1969.

The second day of hearings was led off by Harve Carlson, who spoke
as the Chairman of the Interagency Coordinating Committee (ICC) for the
IBP. Carlson gave a brief résumé of the current state of development of
US/IBP, then emphasized the need for new funding for the major, integrated programs being planned by the U.S. organization:

"By the submission of House Joint Resolution 1240, you and
your committee demonstrate your understanding that the aims of
IBP cannot be met simply by reprogramming on-going research
or by reassigning staff to participate in certain phases of the
program. The major, integrated programs, the largest and most
costly of which is the ecosystem analysis program with its six
separate biome studies, cannot be successful unless additional
funds are forthcoming."

Philip Handler, later to succeed Fred Seitz as President of NAS/NRC,
followed Carlson, speaking as President of the National Science Board. He
led off with the statement: "Unfortunately, I am not as knowledgeable as I
might be with respect to either the programs of the IBP or ecology itself."

This was no news to the ecologists or the supporters of IBP generally,
and it was my impression that he later maintained this level of knowledge
throughout my tenure as Chairman, USNC/IBP. Handler's statement to
the subcommittee was the most negative and half-supportive of any that
was made in any hearing on the IBP, for example:

"We remain, I think, somewhat uncertain as to whether or
not the program in its present format will achieve the actual
objectives to which you look forward.
I am not entirely certain in my own mind that even now the
program can achieve those goals."

If the Daddario subcommittee had let Handler's testimony take
precedence over that of its other witnesses, I would have expected it to
drop the IBP. However, it did not.

The final testimony was an extemporaneous statement by Ed Deevey,
who had accompanied Harve Carlson. Deevey's argument consisted mainly

of a rationalization for NSF's cutting back on requested levels of support and to me seemed counter-productive.

Our Public Affairs Committee had a chance to further promote the concept of ecosystem ecology, when, in July 1968, there was a Joint House-Senate Colloquium to discuss a national policy for the environment. This was jointly sponsored by Miller's committee in the House and the Senate Committee on Interior and Insular Affairs. Our committee was invited to submit a statement. This statement, prepared primarily by Stan Auerbach, was entitled, "The Importance of Ecology and the Study of Ecosystems." Significant points made were:

"Conceptually the ecosystem lends itself to analytical and empirical research as well as the cell, the organ, and other lower levels of biological organization. It includes the processes of circulation, transformation, and accumulation of energy, matter, and nutrient elements through the medium of living things and their activities. Some specific functional processes include photosynthesis, decomposition, herbivory, predation, parasitism, and symbiotic activities. Because it involves processes, interaction, transfer, the ecosystem is at least conceptually amenable to experimental study under field conditions. However, much experimentation requires more sophisticated research and analysis and greater array of research tools than heretofore was deemed essential by ecologists.

"As to outlook, ecologists will have to learn to curb some of their traditional individuality, to learn how to work in large teams harmoniously and effectively, and to develop ways of sharing data. What is needed is nothing less than a new psychology or a new sociology for ecologists. The usual inertia attendant on changes in tradition will be overcome because the intellectual goals require change. Ecology is not different from any other scientific field in this regard: the intellectual values exceed all others."

The next act in this scenario was the preparation of the Program Memorandum requested by Ivan Bennett for the IBP. This was prepared with major input from Richard Oliver, our chief staff person at NAS/NRC. After this exercise, the BOB approved an item of $5 million in the Federal budget for fiscal year 1970.

The Program Memorandum on the IBP that had been requested by OST and mentioned by Ivan Bennett during the hearings on May 1-2, 1968, was to prove to be the key that unlocked substantive Federal support for the IBP. Harve Carlson, as chairman of the Interagency Coordinating Committee, presented an extensive list of questions as a basis for preparation of the memorandum. Our small IBP staff in Washington, especially Dick Oliver, did an outstanding job of responding to these questions.

The results of this study were seemingly satisfactory. In its own report back to OST, NSF stated:

"The National Science Foundation has analyzed the needs and objectives of the United States portion of the International Biological Program (IBP). This analysis was made solely to determine the requirements of a sound IBP program without regards to questions related to the budget process; i.e., no analyses were made of the relative needs of the IBP as compared to those of other scientific programs. It is the conclusion of the National Science Foundation that the U.S. portion of the IBP program will have potential research results of great importance to our national interest.

"The broad objective of the IBP is to provide insight and foresight about the changing relations between man and his environment. The U.S. effort, as an integral part of an international program, is concerned with achieving a better understanding of the impact of (1) the increasing size and needs of human populations, (2) the impact of these increases upon the interconnected ecological systems of the Earth, and (3) the impact of the resulting effects on man. The U.S. effort is organized into major studies called Integrated Research Programs: 12 have been endorsed by the U.S. National Committee for IBP, and 5 are under consideration. Each major study represents a coordinated research effort involving numerous U.S. scientists and often is closely related to IBP efforts in other countries. The scope of most projects is much greater than has been customary in the past, thus making possible a more complete understanding of man and his environment.

"The document describes each of the 17 major studies including relevant details of anticipated benefits, schedules of progress, cost estimates, and priorities. These schedules, priorities, and levels of activity have been determined by examining the scientific promise, feasibility, and status of preparation of each major study. Some of these studies are under way and should be expanded. Others are certain to be ready for significant funding in FY 1970. A few, in the opinion of NSF, need to be modified to increase the participation of social scientists. Efforts will be made to achieve this during the next year.

"Parts of the ongoing activities of various Federal agencies can, with adequate coordinating efforts, be made parts of the Integrated Research Programs. The effect of this varies from agency to agency and from program to program. However, many important parts of these programs cannot be undertaken on an adequate scale, or in some cases cannot be undertaken at all, without additional funds.

"NSF believes that a reasonably comprehensive program will require $8 million of special funds in FY 1970 and a total of $43 million during the period FY 1970 through FY 1973. It is recognized that budgetary pressures may necessitate a reduced program, especially in FY 1970. Only small amounts of non-Federal funds are available. As an aid in the understanding of the effects of different budget levels, higher and lower budget projections are given. With a high level of funding the scope of the studies could be broadened and the intensiveness of research increased. At a low level of funding a programmatic stretch-out will be necessary; benefits will be correspondingly reduced, and some studies may not be started at all."

The word came back that this document had been favorably received and that the Bureau of the Budget (BOB) had approved a line item for IBP in the FY 1970 budget of NSF in the amount of $5 million.

Another indication of NSF support came in late 1968 when Phil Johnson was appointed as Program Director for the IBP in the Foundation, thus establishing a separate identity for the IBP in that agency. One of our 1968 efforts to advance the IBP was mounted at the annual meeting of the American Institute of Biological Sciences (AIBS) at Columbus, Ohio, in September. AIBS had offered IBP a free booth at the annual convention. This was manned by Dick Oliver and others had provided visual demonstration of developing U.S. programs for the IBP. Additionally, I had organized a symposium on the IBP entitled, "Man's Survival in a Changing World." Highlight of the symposium was a talk by Congressman George Miller that was highly supportive of the IBP:

"The IBP appears to us to be the best and most available instrument for stimulating and upgrading the activities of American biologists. Where else can we look to find a concert of interest and organization at the national and international level in new concepts in organic productivity; in our great grasslands and forests; our migrating and high altitude populations all over the world; a last-ditch study of the Eskimos and American Indians; the movement of biological materials in the atmosphere; the subtle effects of pollution in our otherwise stable ecosystems? And do we not need soon a better estimate of the capacity of the earth to support a given population? Where else other than the IBP can we look for a new generation of first-rate ecologists, conservationists and engineers who understand the strategy of nature well enough to provide operations advice to managers of farms, forests, and fisheries — to municipalities sharing a common watershed or atmospheric region — to the design of new population centers and the rebuilding of the ones we have? How long can we tolerate the anguish of uncertainty and indecision based on the lack of adequate data? We need to be told what we can and cannot do — what we should and should not do — backed up by solid scientific fact. This is what the IBP proposes to do."

Fred Smith, Fred Sargent, and I also spoke on this symposium, but we felt that the emphasis on interest of the Federal Government in the IBP expressed by Congressman Miller was of the highest significance.

Strong efforts to advance the IBP moved ahead early in 1969. On March 25, Miller and Daddario introduced H.J.R. 589, calling for Federal support for the IBP. This resolution read:

"Resolved by the Senate and House of Representatives of the United States of America in Congress assembled, that (a) the Congress hereby finds and declares that the international biological program, which was established under the auspices of the International Council of Scientific Unions and the International Union of Biological Sciences and is sponsored in the United States by the National Academy of Sciences and the National Academy of Engineering, deals with one of the most crucial situations to face this or any other civilization — the immediate or near potential of mankind to damage, possibly beyond repair, the earth's ecological system on which all life depends. The Congress further finds and declares that the international biological program provides an immediate and effective means available of meeting this situation, through its stated objectives of increased study and research related to biological productivity and human welfare in a changing world environment.

"(b) The Congress therefore commends and endorses the international biological program and expresses its support of the United States National Committee and the Interagency Coordinating Committee, which together have the responsibility for planning, coordinating, and carrying out the program in the United States.

"(c) In view of the urgency of the problem, the Congress finds and declares that the provision by the United States of adequate financial and other support for the international biological program is a matter of first priority.

"Sec. 2 (a) The Congress calls upon all Federal departments and agencies and other persons and organizations, both public and private, to support and cooperate fully with the international biological program and the activities and goals of the United States National Committee and the Interagency Coordinating Committee.

"(b) For this purpose, the Congress authorizes and requests all Federal departments and agencies having functions of objectives which coincide with or are related to those of the international biological program to obligate or make appropriate transfers of funds to the program from moneys available for such functions or objectives and provide such other support as may be appropriate."

The same resolution was introduced into the Senate on April 3 by Senator Edmund Muskie as S. J. R. 89.

Hearings were called in the House of Representatives on May 6-7, 1969, under Daddario's Subcommittee on Science, Research, and Development.

Again, I led off the testimony as Chairman, USNC/IBP. I presented a detailed statement about the projected research under the IBP that had been prepared by the IBP staff. I followed this with an informal presentation with my main points being:

"Summarizing what I have said, it seems to me, we have a reasonable estimate of what we should spend and can profitably spend on the IBP in this country in fiscal 1970. This is the sum of $15 million, which received a greal deal of consideration by our own committee, by people at the National Science Foundation, and by the Bureau of the Budget.

"I would like to emphasize again that we are thinking in terms of support for the new kinds of integrated research projects. Now, the United States is going to get whatever program it can bring itself to pay for. I think this has to be said. We are going to have some kind of program in the United States, we already have one. However, I would hope that we can get the best that our scientists are capable of providing.

"We have moved to a definite position of world leadership in this international effort, and I would hate to see us have anything but the best we can buy. One would like to cite one illustration of the effect our leadership has.

"The international committee that coordinates the world-wide studies of terrestrial production under the IBP has reclassified its system of indexing projects to take into account the fact that there are the big kind of ecosystem studies that are being done in the United States. Several other countries have reoriented their programs to follow the pattern established by the U.S.

studies. Chile is one. Some of the European countries have also moved in this direction. This is what our leadership means. We are in a position of world leadership, and I hope we can stay there."

Fred Smith followed with a résumé of the progress and funding needs of the various biomes. Fred, too, stressed the need for new money:

"Now, as a second point, I would like to discuss agency support for this program. Because in fact we enjoy very strong support from many areas including the Atomic Energy Commission, the Agricultural Research Service, the Forest Service, et cetera.

"We are using their facilities. We are using their equipment and manpower. I think it would be excellent if these agencies documented this to show the level of their present support of the IBP.

"It would run into several million dollars, but I want to stress as Dr. Blair did this does not constitute part of the money we are looking for. What we are looking for is new money in terms of logistic support."

Fred Milan discussed progress of the Eskimo studies program of the IBP.

Harve Carlson, accompanied by B. H. Ketchum and Philip Johnson, then testified for NSF. Carlson testified that since May 1968, the Foundation had awarded 13 grants in the amount of $1,013,800 for IBP integrated research proposals and that IBP proposals for major programs pending before the Foundation amounted to over $6.4 million.

Carlson testified:

"It is clear the $5 million requested by the Foundation for the IBP will not support the whole program. Certain agencies have been able to support specific research programs closely related to their mission interest."

He concluded his testimony with a favorable evaluation of the progress and promise of the IBP:

"In summary, Mr. Chairman, may I say that the Foundation, following a special study of the IBP completed last fall, is satisfied with the course and progress for the IBP. We have been most gratified by the cooperation of the Agencies which comprise the Interagency Coordinating Committee for IBP.

"Further, the Foundation is confident that both substantive scientific programs and appropriate management are evolving. The Foundation envisions that the output of this innovative research will establish and stimulate team approaches to environmental and ecological sciences at fundamental as well as applied levels.

"We are confident that the results of IBP research will help to provide the necessary platform of basic understandings on which society can base solutions or alternative activities to cope with ecological problems."

In response to questioning, Carlson stated that the incoming (Nixon) administration had approved the $5 million requested in the NSF budget for the IBP.

Opening testimony on May 7 was given by Lee DuBridge, Science Advisor to President Nixon and Director, Office of Science and Technology. His testimony was supportive and was useful in making the necessary distinction between the earlier, simpler IGY and the IBP:

"The IBP is unique in several respects. It represents a major new kind of undertaking for the biological community in that it requires the coordinated efforts of scientists from diverse disciplines who are accustomed to highly individual research projects.

"This is not like the International Geophysical Year, requiring one type of collaboration. This requires more diverse types of activity.

"Further, the Program is addressed to improving, on an unprecedented scale, our understanding of human adaptability to the environment and our knowledge of natural systems. In view of the multiplicity of programs that are directed toward improving man's welfare, it is difficult to estimate the contribution the IBP will make, particularly in those areas concerned with the environment. We should not, of course, expect that a scientific program such as IBP will provide the solutions to environmental management problems. Nevertheless, I believe that the IBP is the type of scientific endeavor that will contribute basic information that can be used to develop methodology for effective management of the environment."

As in other hearings, Daddario was expert in questioning the witnesses and very knowledgeable in his own commentary, for example:

"Dr. DuBridge, I would like to add one further thought: The questions that have been raised here all point not just to how we manage our environmental and biological problems, but to all our programs. I would expect from what you have already said, the work of the IBP, if it continues developing and if properly supported, can go a long way.

"You have already said that this is a program that will not be done in 1 or 2 years but will go on for 5 or 10. It represents a new undertaking that requires coordinated efforts from those of different disciplines. We see multidisciplinary activity growing in a multitude of places where people unaccustomed to working in this way are beginning to. We expect great progress will be made. As we go along, this will fit into the pattern of what we need to do. As this is done, you will be in a better position to make the recommendations and determinations to the administration and the Congress as to what needs to be done.

"My question would then boil down to: Aren't we putting together the necessary elements through which we can come to better determination as to how to better manage our resources in this area than we have in the past?"

Final testimony was given by N. W. Gibbons, Vice-Chairman, Canadian National Committee for the IBP, who described the organization and progress of the Canadian effort.

H.J.R. 589 was reported out favorably on June 11, 1969 by the Committee on Science and Astronautics, and on November 12, 1969 it was considered and passed by the House. Debate in the House was surprisingly long considering the nature and timeliness of the resolution. Strongest opposition came from Congressman Fulton of Pennsylvania, a member of the Subcommittee on Science, Research, and Development and one who had been generally hostile to IBP in the hearings before that subcommittee. Daddario spoke strongly in favor of the IBP:

"Mr. Chairman, the IBP is not merely a series of unrelated programs, tied loosely together. It aims, rather, at an entirely new concept, a concept of totality. Its programs are interrelated and the results will be synthesized into a whole. Its frame of reference is not focused solely upon one discipline. The IBP employs scientists from every discipline; sociologists, physicists, chemists, biologists, oceanographers, and others. It provides in sum, the kind of integrated, interdisciplinary approach not possible in other programs in the past.

"It is this pragmatic, problem-solving approach that has led ecological practitioners in the United States and in over 55 other countries to undertake the complex task of planning and designing the international biological program, which will produce the kind of manpower knowledge, and experience that is essential to the effective management of our environmental resources."

Happily, Daddario's arguments, and those of George Miller were persuasive.

S.J.R. 89, the companion resolution in the Senate, was fortunately referred to the Committee on Labor and Public Welfare, chaired by Senator Ralph Yarborough, a Texas liberal. Again it was my role to lead off the testimony in hearings on February 9, 1970. I had support from Bill Benninghoff, Everett Lee, Jim Neel, and George Van Dyne. Senator Eagleton, who chaired the hearings, led off with a statement to the effect:

"It is an appropriate time as we begin a new decade to discuss this program of ecological research. If, as many of us hope, the 1970's are to be the time when we finally turn the Nation's attention to environmental problems, it is important that we understand something about this system we call the environment and how it works.

"The Congress has dealt in the past with seemingly isolated problems of pollution, food resources, conservation of wild areas, water use, etc.; however, congressional support for ecology—the only scientific discipline which specializes in the total environment—has been conspicuously lacking.

"The International Biological Program—which I should add has been conceived and developed largely by scientists and not government—can be of major significance in helping to make up for the inattention of the past.

"The data which this program can produce about the environment, and of equal importance, about man's adaptability to a changing environment, should form the basis for the programs that we expect will be devised in the years to come."

Endorsements of the IBP by the NSF, Bureau of Budget, Department of State, and Department of Agriculture were read into the record.

In my opening testimony, I presented a rather detailed prepared statement about the IBP that had been prepared by IBP staff. Then I spoke to financial needs:

"It is our understanding that the National Science Foundation will request $7 million for the IBP in its fiscal year 1971 budget. Our own committee in careful deliberations and in very painful deliberations with the PROCOM Committee has estimated that the program that the United States should have in fiscal 1971 would cost $16 million.

"The estimated need for fiscal 1970 was $15 million. Actually we have only $4 million expected in fiscal year 1970 from the NSF plus approximately one half a million dollars from other agency sources.

"So, in fiscal year 1970 we have something like $4.5 million. If you put in services that are hard to put figures on, there is perhaps $5 million for the IBP in fiscal year 1970 as opposed to our considered estimate of $15 million of needed funds.

"For fiscal year 1971 our estimate again is far over what we can expect the NSF to be able to provide in these times of short money. So, I think that this resolution is very important to us in calling the attention of the various interested agencies — and there are many — and several of them are represented in this room today — that have missions which certainly parallel the objectives of the IBP."

I had supportive statements from Jim Neel, Everett Lee, Bill Benninghoff, and George Van Dyne, but, in general, the dialogue was much less extensive than that before the House subcommittee.

Senator Muskie, not a member of the committee, went out of his way to appear before the committee in support of the IBP. He stated in part:

"Mr. Chairman, I appreciate the opportunity to appear before this subcommittee today in support of Senate Joint Resolution 89 or as it will ultimately pass the Congress, House Joint Resolution 589.

"I do so because of my interest to which the chairman has referred, in protecting and enhancing the quality of our environment. I don't suppose anybody is against that objective at this point. It has been well sold.

"But nevertheless we have to come to grips with the very specific task of doing the things that need to be done to achieve that objective. So I am grateful to you, Mr. Chairman, for the interest and concern which you have shown in this program.

"We have learned, I think, that we must take worldwide action to avoid disaster. As an atomic mass becomes critical, our biological communities are also reaching critical stages in development. If we continue to toy with the delicate balance of nature, we will not survive.

"Environmental pollution is simply incompatible with human health and welfare, so we must begin to make changes in our lifestyles. But before we know what kind of changes we should make, we must know where we are in relation to our environment. Before we plan changes, we should be able to predict consequences of carrying out our plans.

"The U.S. participation in the international biological program constitutes, I think, one of our first efforts, perhaps the first, to acquire the knowledge to support environmental planning and management, which must take place not only in a single country, but all around this planet.

"It is not a cure. It will not provide finished answers to our environmental problems. But it is a start toward understanding the relationship between man and his environment and it is a good start."

Final testimony was given by Harve Carlson, with support from John Totter, Director, Division of Biology and Medicine, U.S. AEC, Ernest S. Tierkel, Director, Office of Science and Assistant Surgeon General, Department of HEW, and R. Keith Arnold, Deputy Chief for Research, Forest Service, U.S. Department of Agriculture.

Carlson testified:

"IBP has served as a focusing mechanism to get large-scale integrated ecological research started in this country. In that past, ecological research has typically been conducted through relatively small projects involving one or two senior scientists and a few assistants. A new dimension has been introduced with the advent of multi-investigator, multidisciplinary attacks on significant ecological problems whose solution is beyond the competence of a single investigator working alone. Dr. Van Dyne emphasized this in his statement just a few minutes ago.

"These integrated projects cannot, of course, substitute for the innovative research of imaginative individual scientists. Both approaches must be allowed to flourish simultaneously, and the Foundation intends to continue support of individual ecologists, largely through its general ecology program.

"IBP is also intended to focus attention on aspects of human adaptability to environmental stress, including the stresses of harsh climates and rapidly changing social conditions. As with the studies of ecological processes in nonhuman communities, IBP research in human ecology emphasizes an integrated attack on complex problems by teams of competent investigators from several disciplines."

S.J.R. 89 was passed by the Senate on August 3, 1970, and was signed into Public Law No. 91-438 by President Nixon on October 7. The IBP now had the official blessing of the Federal government. Of more importance than this official endorsement was the effort to obtain adequate Federal funds so that the big ecosystem-oriented studies could go forward.

When time came for the House Committee on Science and Astronautics to hold authorization hearings on the NSF FY/70 budget in the Spring of 1969, Chairman Miller invited the IBP to participate in that portion having to do with the IBP. The word soon got around that this was annoying to some NSF bureaucrats, but there was nothing that they could do about it. I will always appreciate the support we had from Chairman Miller. As we were sitting in the hearing room, waiting for the precedings to start, Miller came into the room, came down from the dais to where I was sitting and greeted me warmly and personally. I don't think the message of support was lost on the NSF men sitting nearby.

Stan Auerbach and Bill Laughlin helped me to present the case for funding the IBP research. My arguments stressed improbability that agencies other than NSF were going to contribute major funding to the IBP:

> "The NSF and our committee concurred in feeling that something in the order of $15 million would be necessary to really launch a respectable IBP effort in the United States in 1970. I think we are very close together on this.
>
> "We have within our committee quite definite plans as to where every dollar would go if we had $15 million available in fiscal year 1970.
>
> "Now I do see some problems with respect to funding. Specifically, the budget study recommended a $5 million 'line item' in the NSF budget, and predicted that an additional $10 million could be found from other agencies; that is, from the budgets of other Federal agencies. I would like to separate these for purposes of discussion.
>
> "First, I will speak to the $5 million budget. Assuming that we do have $5 million in the NSF budget, there is a problem posed by the present limitation on spending, which has been imposed on the NSF, and by the NSF on universities.
>
> "I can see that we don't have $5 million, even if we do have $5 million in the budget. I think the average restriction on expenditures is roughly 25 percent. This means that, if we got this $5 million in the NSF budget, and operated under the restrictions which are presently in force, and which many people are projecting into 1970, we would have three-fourths of this, or 3.7 of this, instead of $5 million.
>
> "This is part of the problem.

"I am not suggesting that IBP receive special treatment, I am just pointing out the facts.

"Secondly, I see no real probability that $10 million could be found in other agencies' budgets.

"The mechanism whereby the IBP has been at least theoretically coordinated at the Federal level is the Interagency Coordinating Committee (ICC). The reasons why this committee has been relatively ineffective are quite simple. The people who have represented the agencies on the ICC were at the level where they could commit their agency to nothing. If we depend upon the kind of ICC that we have had up until now, I think it is wishful thinking to think we will round up $10 million of support from the other agencies. Therefore, I feel that the projection of $15 million is unreal, and I feel that if some mechanism isn't developed to make this $15 million really available then we are going to be badly set back in getting our IBP into operation.

"I would say that, speaking first to 1970, if we really had $15 million we could put into—using some priorities—either really full-scale operation or pilot operation the entire program that has been planned by our committee. At a level comparable to this, and with some expansion, we could do a very respectable, and, in fact, scientifically outstanding, series of programs under the IBP.

"Perhaps in this context I should emphasize something that I have not yet said. The emphasis in the IBP program of the United States has been on our integrated research programs (IRP's). These are coordinated research programs that would not be in existence without the IBP.

"In some countries participation in the IBP has consisted of labeling what is already on-going research and saying, 'Yes, we are in the IBP.' You are concerned about the environment, man's relation to a degraded environment. The IBP has meshed with this concern, and I think this would have come eventually whether the IBP was here or not, but the IBP has been a catalyst. I am talking about the studies that are essential to an understanding of what we are doing to our environment. Ecologists have been talking about this for a long time, but until the IBP came along and catalyzed this, nobody had the courage to do it. Initial experiences with the first ecosystem analysis project—Grasslands— are enough to encourage us in the approach we have taken. We are able to get people to work together toward understanding how an ecosystem works, and this has never been done before."

Stan Auerbach stressed the high degree of interest among American scientists in becoming involved in the Integrated Research Programs of US/IBP. He also emphasized the societal needs for the ecosystem approach that was being planned under the IBP:

"Mr. Chairman, I would like to make two comments. One of the gratifying developments that has occurred during the past year has been the increased willingness of the scientists who are involved in the IBP to cooperate within a central coordinating structure. We have set up a series of internal committees which review each proposed program before we recommend it to a funding agency. These reviews have been, and still are actively being carried out within the integrated research program structure; and, in many cases, the proposals for programs are being sent back to the groups and for reworking in order to meet the criteria of an integrated research program. We find that their response to this type of suggestion has been most gratifying.

"I wish to make a point of this, Mr. Chairman, because I believe there was some concern felt last year during the hearings on IBP whether ecologists and related scientists would work in a cooperative type program. The evidence to date strongly suggests that they will.

"Second, there are a number of further technological threats to our ecosystems developing, that place a rather urgent priority for early initiation of our Biome programs. The most recent is the Alaskan oil strike, estimated to be 300 billion potential barrels of oil. This find already poses a threat to the tundra ecosystems — especially since we absolutely lack information on the recovery of damaged tundra. Furthermore, we have neither State nor Federal regulations to deal with this situation, and the agencies responsible for development of regulations lack the information on tundra damage and recovery necessary in order to develop regulatory structures which are necessary under these circumstances.

"Dr. Blair has mentioned our concern about the tropics. In the Analysis of Ecosystems Project we have been looking at a number of possible sites in the tropical area where South American scientists could converge and work together. We have in mind a site in South America where it would be possible for the scientists from the different countries in South America to work together in tropical ecosystem problems.

"In the eastern United States we have four major technological problems developing. One is the increased population. Two is the increased demand for living space and a concomitant increased demand for recreation.

"The third is the increased pollution of the eastern region, and its impact on the ecosystems of this region. The fourth is the demand for timber products. The cost of timber is rising rapidly, and in response to this demand we see two developments and we are worrying about their long-term impact on our ecosystems.

"The first is a shift to the increased planting of hybrid trees, and the second is the increased use of fertilization to speed timber growth. The latter is of special concern since much of our eastern water supply is derived from forested watersheds. Fertilization of these watersheds could lead to a compounding of our water pollution problems in the eastern deciduous forest region.

"In the desert biome we recognize that the desert has increased potential for human habitation, as well as for additional grazing. Both represent ill-defined threats to these regions. Consequently we are concerned that in order to meet the needs of the technological assessments that are required, we should start to use the systems techniques that are now becoming increasingly available. Furthermore, ecologists are becoming more and more interested in using these techniques to evaluate these technological threats to our ecosystems. Because of these technological threats to our ecosystems, we believe that it is to everybody's advantage to start all six biome programs as soon as possible or feasible."

Bill Laughlin pointed out the urgency of getting the Eskimo study into high gear.

"I might start with the Prudhoe Bay oil strike previously mentioned. This is going to move a group of Anaktuvak Eskimos that were challenged by strontium 90. They eat caribou that eats moss that picks up strontium 90. They radiate, and so they have been the subject of study for some time because of the high loads they carry. They can tolerate that much, apparently, and therefore, we know we can. The most serious threat is the road and pipeline going through Anaktuvak Pass. Many of these Eskimos will go to work for the pipelines. This means two things. They are going to lose their hunting ability and they will get fat like other Eskimos that do not get out and hunt. Further, they will marry outsiders, and the genetic structure will change, and so forth. This is an example of what has been happening in the Arctic. Eskimos were an ideal group for international cooperation, because there are Eskimos in Alaska, Greenland, Russia, and Canada. There are Aleuts in Russia as well as Alaska, and there has already been some degree of scientific cooperation over a number of years."

Our presentations were well received, and the IBP "line item" was in the NSF budget from FY/70 to the end of the IBP, reaching an annual peak of $10 million. In retrospect, I am convinced that the IBP would have been a paper exercise, as it was in so many nominally participating countries, if it had not been for the enthusiastic support given by Congress-man Miller and Daddario, Senators Yarborough, Symington, and Muskie, and agency staffers like Dick Carpenter, Freeman Quimby, and Marcia Carlin. Sharon Friedman, who was assigned by NAS/NCR to full time public relations for IBP, deserves much credit. Above all, Dick Oliver, who ran our Washington office, was always able to provide position papers on short notice, was usually able to set up key appointments on short notice, and could provide back-up for me at these appointments even though some of the higher echelons at NAS/NCR frowned on such contacts.

In May 1971, I had an excellent opportunity to call attention to the IBP during Congressional Hearings on a proposed National En-vironmental Laboratories bill. In February 1969, Senators Howard Baker and Edmund Muskie had co-sponsored S. 3410 to establish a system of Natural Environmental Laboratories (NEL's), basing their proposal on the plan worked out by an *ad hoc* committee that had met at Oak Ridge National Laboratory in December 1969. No action was taken in the 91st Congress, but essentially the same bill was introduced into the 92nd Congress, as S. 1113 on March 4, 1971, with an additional 27 sponsors.

Extensive hearings on S. 1113 were held by the Subcommittee on air and water pollution of the Senate Committee on Public Works from April 28 through May 6, 1971. A wide spectrum of agency people and of environmentalists was asked to testify. I was called on as Chairman, USNC/IBP. My prepared statement for the record was largely a progress report on US/IBP. I closed my oral statement with:

"I would close with emphasizing that there is a very definite interface between the NEL's, IAIE [Interamerican Institute of Ecology, later renamed the Institute of Ecology], and a very definite contribution from the IBP that may give us a running start in developing the kind of objectives that are expressed in this proposed legislation."

Congressman Miller provided one other major push for the IBP and for the things it was trying to accomplish. One day, Dick Carpenter and I were in Chairman Miller's office briefing him on the progress of the IBP. In the course of the conversation, the idea emerged that it would be desirable to hold a joint House-Senate colloquium on international envi-ronmental science, and in so doing highlight the IBP. This would be only the second such colloquium ever held in the history of the U.S. Congress, the first being held in 1968 and also on the subject of the environment. Chairman Miller like the idea and found similar sentiment on the part of Senator Warren Magnuson, Chairman of the Senate Committee on Commerce. The result was that a two-day colloquium was organized for May 25 and 26, 1971.

Magnuson was one of the strong environmentalist voices in the Senate, and on April 27, 1970, had introduced S.R. 399 to create a World Environmental Institute. He later promoted this same idea at the UN Conference on the Human Environment at Stockholm in June 1972, but without success.

A prestigious international group of environmentally involved participants was invited. Tom Malone, Vice-President of the International Council of Scientific Unions, was named Rapporteur. After opening remarks by George Miller and Warren Magnuson, statements were heard from Maurice Strong, Secretary-General U.N. (Stockholm) Conference on the Human Environment; Russell Train, Chairman, U.S. Council on Environmental Quality (CEQ); Peter Walker, Secretary of State for the Environment, Great Britain; Christian Herter, Jr., Special Assistant to the Secretary of State for Environmental Affairs; Francesco di Castri, Vice-President of ICSU's Scientific Committee on Problems of the Environment (SCOPE); Tom Malone; Kwan Sai Kheong, Permanent Secretary, Ministry of Education, Republic of Singapore; Athelstan Spilhaus, Woodrow Wilson Center for International Scholars; Roger Revelle, Director, Center for Population Studies, Harvard University; and my predecessor as Chairman USNC/IBP, Herman Pollack, Assistant Secretary of State for Science and Technology.

From the IBP, we had myself; B. R. Seshachar, President, Indian National Science Academy and Chairman, Indian Committee for the IBP; and Bengt Lundholm, Secretary, Ecological Research Committee, Swedish Natural Science Research Council.

In my own testimony, I first gave well-deserved credit to Chairman Miller:

"A further pleasure comes from participating with Chairman Miller, whom we regard as somewhat of a godfather, along with his committee, for the International Biological Program efforts in the United States.

"I think that the efforts they made to bring this program to the attention of the Congress and the public a few years ago were very instrumental in the success that we have had in the International Biological Program. So, I particularly enjoy participating here with the chairman."

I then proceeded to stress the hemispheric nature of the environmental problems with which we are confronted:

"I would like to put my remarks this morning in the context of the spaceship concept of the world which now is a pretty well-accepted concept, and I would like to point out two aspects of this spacecraft.

"One, of course, is the enormous complexity of the in-terrelationships of the functions and events that take place on this spaceship. We have come to recognize the fact that perturbations of this system have effects in fantastically distant parts of the system.

"Second, in keeping with this concept of the spaceship earth, we must recognize the fact that the resources of this planet are absolutely finite. I think all of our considerations must take into account these two facts: the interrelatedness of events and the finiteness of the world's resources.

"This being the case, looking at the spaceship from the viewpoint of one of the so-called developed countries, we must consider that, as world leaders in the ecological science, we have an obligation not only to the other parts of this planet, but to ourselves to do all we can to promote development of competence in environmental science in other countries, in the lesser developed countries. This is not solely in their interest, but also in our own interest, because we are a part of a complex system and one part of the system affects other parts of the system.

"Now, I would like to address my remarks, largely, this morning to the part of this planet outside of the United States that I know best. This is Latin America, particularly South America. Perhaps, when one says South America, if you have not been there, you tend to think of a uniform area. Such is far from the case.

"Now I would like to emphasize the ecological diversity that must be taken into account when one starts thinking about the rational development and management of the resources of this vast continent to the south of us.

"I do feel that I have some competence to speak in terms of the needs and in terms of the present competence of the Latin American countries to solve their own problems without assistance from others. It may be presumptive to say this with my good friend, Dr. di Castri, across the way from me, but since he and I have worked very closely together, I think he will probably agree with most of what I say, and if he doesn't, I am sure he will say so.

"Actually, if we think in terms of trying to manage the en-vironment in South America, we have to think in terms of three huge ecological divisions. There are subdivisions of these, but basically we think in terms of three huge regions. One of these is the tremendous area of tropical forest in the basin of the Amazon and Orinoco, with patches elsewhere. This is the largest area of tropical forest in the world. It has probably felt less of man's impact so far than any other comparable area of the world.

"Because of this and because of its lushness in terms of production of organic materials, the tropical forest, and especially the neotropical forest, is the continual target of schemes for exploitation. Yet, it is one of the ecological systems of the world about which we know probably the least. There is not time to go into the details, but there are peculiar features of the tropical forest such as its enormous biological diversity and its complexity as an ecological system.

"What we do know is that if you destroy the tropical forest you have upset a very complex ecological system that does not quickly recover. Consequently, many efforts to develop agriculture in the tropical forest lands have been failures. Yet we have very little scientific knowledge of the tropical forest as a system. In fact, one quotation that I picked up points out that there are 10 experiment stations in all of the Amazon tropical forest area, and these have each essentially one scientist with at least a high school education.

"This is the kind of competence we have in managing this huge forest. I think one of the greatest challenges to the ecologists of the New World is the development of the knowledge, the know-how to develop the tropical forest without destroying it.

"The second area is the Andean Highlands. It is an area which in pre-Columbian days supported more of the indigenous population than all the remaining area of South America combined. It is a very harsh area. We know how man uses the environment there. It would be a logical place in which social scientists and hard scientists could get together to look at man interacting with his environment and to produce knowledge for the better use of that system.

"The third area is the very large area of drylands in southern and eastern South America, the 'monte' and 'chaco' of Argentina, the 'cerrado' and 'caatinga' in Brazil. Rainfall is deficient. There are great pressures, particularly in Argentina, to develop this land to take care of the needs of their growing population. Here man has interacted greatly with the environment, particularly through using goats as a domesticated animal, and there are tremendous effects.

"These are the three major areas that need to be looked at in terms of how they can be best managed and best developed without making the mistakes that most of the developed countries have made in their past history. I think it is an obligation of the developed countries to point out their own mistakes as exemplary of what not to do and to interact with these countries as they develop.

"Now, what are the capabilities of the South American countries to solve their own problems? I should expand here a bit on what I mean by capabilities. I am thinking in terms of the kind of sophisticated, multidisciplinary studies that we have found successful in the International Biological Program, particularly here in the United States, where we look at the whole ecological system, with many participants, and with the ultimate objective of a predictive model of that system so it can be used by our decision-makers in the management process. This is our aim and objective. This is what I mean when I say ecological capabilities.

"As Dr. di Castri well knows, we have in Latin America many scientists who can fit into such a pattern. What we do not find there in very many places is the ability to put all of this together into a sophisticated ecosystem-type study such as we have developed in the International Biological Program."

Seshachar, not unexpectedly, spoke almost entirely to the almost insolvable environmental problems of this country.

"The problems of environment of a country like India are not only different from those of highly developed countries but some of them are peculiar to it.

"India has an area of little more than a million square miles. Its population, according to the latest census which came through only a few weeks ago is 550 million—a seventh of the world's population. Eighty percent of these people live in small settlements called villages which number 600,000. Some of the largest urban centers of the world are in India—Calcutta with 7.5 million people, Bombay with 6 million, Delhi with 3.7 million.

"With a history, culture, and civilization which are among the oldest, India is one of the world's great human centers and stands as a mighty challenge not only to the ingenuity of man but also to his dignity. In fact, it is this ancient civilization that has acted as a social and cultural deterrent to economic development. That is the problem of India. But the compulsive needs of science and technology cannot be ignored and India should prepare and is preparing for them.

"Mr. Chairman, in the little time that I have at my disposal, I can do little more than briefly touch on some aspects of India's environmental problems which might be of interest to this group, and which, hopefully, would constitute areas for further study.

"The first problem is our population. I don't have to make an apology for it because among the first remarks that were made in this room yesterday by Senator Magnuson there was a reference to population. When one talks of the population increase, he asks four or five questions:

"First, has it threatened total food production and its distribution in a damaging way?

"Two, has it thrown pressures on resources and services and caused disruption of these resources?

"Three, has it impeded, retarded, or arrested economic growth?

"Four, has it led to an unrelievable pollution problem?

"And, finally, has it led to a sharp disproportion of working and dependent population?

"Population is not merely India's problem. It is a problem of many parts of the world. It is a question of how many of these criteria are met. In several countries, some one or the other of these criteria are relevant. In India, all of them are relevant. So I need not, to a highly informed group such as this, emphasize the highly intimate relationship that exists between population increases and its effect on environment, that this relationship is even more complex, and poverty accompanies population increase is nowhere clearer than in India. But there are other features of the Indian situation which are of great interest.

"The disparities between urban and rural settlements are sharper and more acute. The cities of India approach, and some ways surpass, the conditions obtaining in some of your urban complexes in the deterioration of their environment, while our 600,000 villages are so primitive that some of them have not changed over centuries. You have an urbanization of your villages—if you have villages at all. We have a ruralization of our cities. In fact, the term village cannot be applied any more to your smaller settlements, except perhaps euphemistically. Extensive industrialization and effective communication have led to the disappearance of the village of the American landscape and it is only a matter of time before it disappears from the American dictionary.

"The worst of both the worlds is therefore characteristic of the urban and rural settlements of a developing country like India. Overcrowding, pressures on services, problems of waste disposal, pollution—all are on an intensified scale in the city. Primitive conditions of living, poverty, absence of protected water supply, lack of essentials for healthful living—these are the problems of our rural areas."

Lundholm stressed the need for international exchange of environmental information and mentioned the planning for a global monitoring system. He also spoke favorably of the U.S. biome studies in the IBP.

"During the last 2 years remarkable things have happened within this program. It started with the biome studies in the U.S.A., which now have developed into a quite new type of international scientific cooperation. It is a cooperation directly between scientists and scientific institutions at the grassroots level in order to find out how the grassroots function. It is a cooperation and international planning in order to get results, which all the participating scientists regard as essential in order to understand the processes of the biosphere. The research is done without bureaucratic overhead. It is fundamental research which will be the base for the understanding of productivity of the earth. IBP will soon be completed, but the biome studies have just started. I think it is very important to keep this research intact and nurse the spirit of international cooperation without political interference."

This whole proceeding was an interesting introduction for me to the way the game is played on Capitol Hill. The sessions were held in the Old Supreme Court Chambers in the U.S. Capitol. Senator Magnuson kept an alert eye on the corridors. Occasionally a stalwart, like Hubert Humphrey, Hugh Scott, William Fulbright, or Teddy Kennedy would be snared and brought into the Colloquium—Magnuson would interrupt the session to introduce each and then give each a chance to say a few brief words. In this way we got a very prestigious list of participants.

In our efforts to align the IBP efforts in the United States with the problems of the times, contacts were made with various Washington agencies, including internationally-oriented ones. Through efforts by Dick Oliver, a seminar with 50 to 60 key individuals at the World Bank was planned for May 23, 1969. The title of the seminar was, "Environmental Effects of Developmental Projects." Stan Auerbach, Bill Laughlin, and I presented papers. Possibly as a result of this presentation, the World Bank in mid-1970 established an ecology office, employing Jim Lee, formerly of HEW and for a time a Vice-Chairman of USNC/IBP.

During this general period, we made every effort we could to bring the IBP to the attention of the general public and the business community as well as to that of government. Primarily through the efforts of Sharon Friedman, a high-level conference, jointly sponsored by USNC/IBP and the Public Affairs Council (PAC), representing more than 200 leading U.S. Corporations with public affairs programs, was organized in Washington for February 18-20, 1970. The theme of this conference was, "Environment: The Quest for Quality."

Keynote speaker was Robert H. Finch, Secretary of HEW. Banquet speaker was Arthur Godfrey, who gave a moving statement about environmental problems.

Bioscience magazine (May 1, 1970) gave the conference a favorable review:

> "The conference was an unqualified success in one major respect — it brought individuals with widely divergent viewpoints and backgrounds together in the hopes of solving a common problem. If they disagreed as to the causes and extent of environmental degradation or the methods needed to combat it, they were at least willing to communicate their disagreements openly in an atmosphere of mutual cooperation. There is now reason to hope for a continued multidisciplinary approach if time does not run out."

During the period of 1968-69, I tried on every trip to Washington to broaden the contacts between the US/IBP and potentially interested organizations and agencies. Prior to each trip, I would ask Dick Oliver of our IBP staff to arrange a schedule of meetings with key Washington people either before or after the meeting for which the trip was being made. In 1968 I made a total of 18 trips to Washington and spent a total of 41 days there. In this period I was able to discuss the IBP with a diversity of interested persons. Several meetings were held with Jesse Perkinson and others in the Organization of American States (OAS) in hopes of obtaining support for our efforts to develop cooperative projects with Latin American countries. The talks were very friendly; the results were nil. We met with John Totter and others at AEC, where we found a very friendly response to the IBP. We talked with administrators in the Department of Health, Education, and Welfare in an effort to elicit support for the Human Adaptability component of US/IBP. We talked with Harrison Brown, then Foreign Secretary of NAS/NRC. Meetings were arranged with Fred Singer and others in the Department of the Interior and with administrators in the State Department. We met with key people at NASA to discuss remote sensing techniques as they might relate to the IBP research. Contacts were also developed or firmed up with various nongovernmental organizations. I met with Russell Train and John Milton of the Conservation Foundation and lunched with Caryl Haskins of the Carnegie Institution. Other contacts were with the National Wildlife Foundation and with John Olive, Director of AIBS. This same pattern of exposing the IBP to all who might be interested was continued through 1969.

In mid-1968 we published a layman's language account of the US/IBP under the title of "Man's Survival in a Changing World." This was mainly the brain child of Sharon Friedman, who conceived and authored this publication. Some 60,000 copies were distributed, and it was later reissued. This document described the U.S. projects under the IBP in laymen's terms. A colorful cover was copied from one of my kodachrome slides of the endangered Brazilian tree, *Araucaria*.

Also, early in 1968 we began publication of the "Interamerican News" in English, Spanish, and Portuguese versions. This highlighted the proposed IBP programs in the Western Hemisphere nations. This publication was continued through the summer of 1971, when it was discontinued because of conflicts between our office and the staff of the Division of Biology and Agriculture at NAS/NRC. The last issue, in mid-summer of 1971, produced a real hassle between our office and a senior staffer in the Division of Biology and Agriculture, NAS/NRC. We planned to run a photograph of Hal Mooney of the Mediterranean Scrub subprogram being greeted by the late President Salvador Allende of Chile, an avowed Communist. The senior staffer objected, we objected to his objection, and we printed the photograph. Needless to say, the whole thing seemed ridiculous since the IBP was an international program without political connotations, with such disparate participants as South Africa and Communist Bloc Countries such as Russia, Czechoslovakia, Hungary, etc.

4

The International Scene

The IBP was, of course, as the name implies, designed to be a global cooperative effort. As a global scientific effort, it is only fair to describe its successes as less than spectacular. Only a few countries other than the United States were able to mount new, really innovative research under the banner of the IBP and to secure funding for such new programs. The common pattern was one of identifying on-going research as IBP. The lesser developed countries of the Southern Hemisphere identified themselves with the IBP, but for the most part did little more. Very few really binational or multinational cooperative projects were ever able to get off the ground under the IBP. Nevertheless, the IBP did have its successes at the international level. It brought scientists concerned with world environment and man's interaction with the environment into a higher level of intercommunication than had ever been accomplished. In ancillary activities the IBP was able to significantly influence other international developments, such as ICSU's Scientific Committee on Problems of the Environment, the U.N. Conference on the Human Environment and its subsequent United Nations Environmental Program, and UNESCO's MAB Program.

I found myself involved in three major activities on the international scene after becoming chairman of USNC/IBP and subsequently being elected one of the four Vice-Presidents of the Special Committee for the IBP (SCIBP). One of these was an ancillary effort to develop a global system for environmental monitoring. One was the effort to extend the official life of the IBP for two years, to mid-1974, because of the late start by countries like the United States, with its complex biome studies and with its late provision of funding for these undertakings. The third effort was one to ensure a successor to SCIBP as an unbrella for international cooperation in ecosystem research at the nongovernmental level.

The concept of global monitoring of the environment was interjected into the IBP at the III General Assembly in Varna, Bulgaria. I had to leave early because of plane connections, but I later received documentation of the last session of the III General Assembly of IBP from Bengt Lundholm. Carl-Göran Hedén of Sweden introduced the subject of global baseline monitoring:

> "As mentioned yesterday there have been some informal discussions about the need for some strategically-located background

stations equipped to measure the biological effects of the regional and possibly global spread of pollution. Fifty years from now we may be criticised most emphatically if we fail to use some of the carefully-selected IBP stations and areas to plot such biological baselines which might illustrate the effect of parameters like turbidity, acidity, the level of oxygen and of carbon dioxide and so on. Dr. Lundholm of the Swedish delegation has repeatedly reminded me of this need, expressed at OECD and by several national governments, and I have been convinced that IBP should consider *what* should be measured in this context and *how* it could be done. Should bio-medical effects and effects on materials be measured, should pesticides be included, which heavy metal ions should be determined, etc.?

"It would be most unfortunate if we forget that all IBP stations and areas are linked by the air above them both to soil and to water pollution. Recent data on mercury distribution and rain acidity indicate that we should treat air, water, and soil pollution as a uniform threat which may penetrate very remote areas indeed. However the field-instrumentation needed must obviously be standardized internationally and the target substances defined.

"I wonder if it would not be wise to appoint an intersectional working group charged with defining the air, water, and soil parameters which should be monitored from the 'pollution baseline' point of view and proposing where and how they should be determined. Perhaps such a group could also consider the case made by Prof. Duvignaud for the statistical comparison of climatological methods. As indicated by the sectional reports presented yesterday not only PT but also CT, PF, and PM might however contribute materially to the activities of such a working group."

As I later learned from Lundholm, Jean Baer of Switzerland and President of SCIBP, objected to this diversion of IBP efforts, but a rump session went ahead and set up a global monitoring committee with Lundholm as chairman and myself and N. N. Smirnov of the U.S.S.R. as members.

As summarized in a June 1968 memorandum by Barton Worthington, who chaired the rump session on monitoring at Varna, the main points made were:

"(a) Definition of problem and reasons for a global network of base-line stations to monitor over a long time period changes in biological communities and changes in the abiotic environment.

"(b) Brief review of international and national stations which exist at present and suggest where new stations need to be established; also the need for stations of 'minimum and maximum' change, 'before and after' studies, etc.

"(c) Reference will be made to the type of biological and abiotic parameters that should be measured over a time series, at the same time stressing the dangers involved in predicting what environmental factors are, and will be, of critical importance of biological communities. Amongst the parameters that might be monitored are the quality of precipitation, carbon dioxide and oxygen content of air, heavy metals, hydrocarbons, biological indicators, selected tissues from plants, animals, and humans.

"(d) The need for a *museum* or *biological bank* for selected materials so that back reference can be made with the use of monitoring methods which are not yet apparent."

Lundholm had promised at Varna that a paper on baseline stations would be produced in consultation with myself and Smirnov. Such a paper was put out in very preliminary form by the Swedish Ecological Research Committee of their Natural Science Research Council in December 1968.

Lundholm, Smirnov, and I agreed that our first steps toward a report would be to convene panels of experts in our respective countries. We would later collate the results of these meetings into a committee report and recommendations to SCIBP.

Implementation of these plans in the United States began in December 1968, when at the AAAS meetings in Dallas on December 26-29 I discussed strategy with Bengt Lundholm, Glenn Hilst, Fred Singer, and Don Farner. It was agreed to convene a meeting on global monitoring in Washington in early spring, with invitations to Federal agencies and other agencies interested in all aspects of global environmental monitoring. It was also agreed that there would be an open meeting of our three-man SCIBP committee on monitoring at Johannesburg, South Africa, in July 1969, in connection with a meeting of The International Union of Pure and Applied Chemistry.

The Washington conference was held on April 18, 1969 under auspices of USNC/IBP, and under the label of "Need for a Global Network for Environmental Monitoring." This was a working conference with four work groups organized around monitoring of soil (chaired by Ted Byerly), water (chaired by Fred Singer), atmosphere (chaired by William Kellogg), and surveillance (jointly chaired by Robert Citron and Dale Jenkins). The four work groups included 43 people, mostly from Federal agencies, plus six NAS staffers. Sören Svensson of the University of Lund was sent by the Swedish Ecological Research Committee to provide liaison with their efforts. Michel Batisse of UNESCO participated and discussed

UNESCO interest in environmental monitoring. The reports produced by these work groups provided a beginning toward the report eventually produced by the U.S. effort toward designing a global network for environmental monitoring. However, the one group that was under-represented was the biologists. Consequently it was decided to convene on May 21 a small group representing a spectrum of biological expertise. In addition to myself, Dale Jenkins (Smithsonian), Ray Johnson (Interior), the invitees included Denzel Ferguson (pesticides in vertebrate populations), Hal Mooney (plant ecology), Arch Park (Department of Agriculture and remote sensing specialist), Ruth Patrick (monitoring pollution in aquatic systems), Tom Mabry (plant biochemistry), and David Reichle (ecosystem ecology).

Our U.S. group produced a brief interim report on June 3, 1969 entitled, "Need for a Global Network for Environmental Monitoring." This was formally transmitted to Barton Worthington for delivery to the SCIBP Committee on Monitoring and was distributed to various Washington agencies and legislators. The general conclusions of the report were:

"(1)   A Global Network for Environmental Monitoring is feasible and desirable.

"(2)   Technology exists to monitor both the physical/chemical environmental characteristics and the biological functions and perturbations of selected organisms and populations."

The general components of the proposed network were defined:

"A reliable Environmental Monitoring Network should include *baseline* stations, *impact* stations, and provision for the use of *indicator* and *sentinel* organisms. These terms are defined as follows:

Baseline station—a station capable of monitoring both physical/chemical characteristics and biological perturbations and functions. It should be: (1) secure and permanent, (2) adequate for the acquisition of baseline measurements—e.g., a national park or agricultural research station, (3) stable with respect to exploitation—i.e., not fragile, (4) located within a major ecosystem of the world, and include both terrestrial and aquatic locations, (5) capable of establishing and monitoring a data bank, and (6) capable of drawing on historical data.

Impact station—a station that is or may be subject to major environmental modification. A reasonable number should be located in areas of monocultures such as corn, cotton, reforestation for industrial harvest, etc. Impact stations should be capable of feeding data into baseline stations.

Indicator organism — an organism globally or regionally distributed and which reflects measurable or observable biological change as an integrated function of environmental changes.

Sentinel organism — an organism purposely distributed globally or regionally which reflects measurable or observable biological change as an integrated function of environmental changes."

At the same time that the IBP committee on Global Monitoring was getting started, ICSU mounted a somewhat parallel effort. At its 12th General Assembly in Paris, September 28 to October 2, 1968, ICSU established an *ad hoc* committee on Problems of Human Environment, with three members to be nominated by the International Union of Geodesy and Geophysics (IUGG) and three by the International Union of Biological Sciences (IUBS). IUBS members were Bengt Lundholm, Don Farner of the United States, and Eric Smith of the U.K. IUGG members were K. Grasshof of West Germany, W. Laszloffy of Hungary, and Ted Munn of Canada. At the first meeting of this committee in Washington, D.C. on March 1-2, 1969, Lundholm was named chairman. A second meeting was held in Stockholm on June 10-11. In September this committee produced a report that had significant recommendations for ICSU:

"The *ad hoc* Committee recommends that ICSU *set up a Scientific Committee on Problems of the Environment (SCOPE)*, which would bring together the various disciplines, and which, through its Commissions, would be responsible for the promotion of environmental monitoring, evaluation of the effects of environmental disturbances, simulation modelling and predictions, and the study of the social effects of man-made change in the environment.

"It is recommended that SCOPE be provided with a secretariat which should be developed into an International Centre for the Environment.

"The *ad hoc* Committee recommends that SCOPE and the International Centre for the Environment cooperate fully with other groups, outside of ICSU, including all involved UN Agencies, regional intergovernmental bodies, and international non-governmental bodies."

The first of these three recommendations was later implemented, and SCOPE became the most significant, and almost only, international entity at the nongovernmental level in the environmental area, except SCIBP.

In the meantime, in the summer of 1969 a seminar, jointly sponsored by the ICSU committee and our IBP committee, was held at Johannesburg on July 19, with members of the IUPAC conference invited to attend. Smirnov from our IBP committee was unable to make the trip. Lundholm and I spoke, as did Fred Singer.

Soon after my return from Johannesburg I received from Lundholm and from Smirnov a copy of the Russian report entitled, "The Opinion of the U.S.S.R. Committee of Biologists on the Global Network of Baseline Biological Stations." The Russian report was consistent with our own report, although less detailed in its recommendations. In his cover letter to me Smirnov included the statement, "I am sending congratulations on the U.S. big success in conquering the moon," referring to our astronauts' successful landing there.

On July 31 Fred Sargent sent a copy of our interim report from the U.S. group to the Secretary-General of the World Meteorological Organization. On August 5 Stan Auerbach wrote me a thoughtful letter on the subject of environmental monitoring. Among his thoughts were:

> "I thought I would pass along some further thoughts on the global monitoring network concept that I have culled during the past couple of days. First of all, I had a talk with Fred Smith, and I think I have arrived at the nub of our differences. Fred feels that the IBP should be concerned with research on problems; that monitoring has nothing to do with solving problems in a research sense. I pointed out to him that this is a basic philosophical difference and that I did not necessarily agree with him, because I felt that if you establish a worldwide network of research on problems, one of its major contributions would be to help man a set of stations which would help us keep abreast of the most important problem of all; namely, what is occurring to the environment on a global scale. Global activities are beyond the scope of individual researchers and the situation is so serious that we cannot afford to indulge in problem solving only on the individual researcher basis. On that basis of philosophical difference we ended the discussion."

The next stage of this scenario on environmental monitoring came at the July 1969 meeting of USNC/IBP when I was instructed to send to SCIBP President Jean Baer and SCIBP Executive Secretary Barton Worthington the following resolution:

> "The Executive Committee for U.S. participation in IBP during July meeting resolved and recommended as follows: 'Recognizing that the IBP already is monitoring certain environmental parameters relating to ecosystems, Appreciating that

surveillance of selected biological phenomena is a critical component of any global environmental monitoring network, and that
various physical and biological networks can greatly benefit from
mutual exchange of data, and Noting that the SCIBP has formed
a working group which is reporting on the feasibility and possible
design of a global biological monitoring network, the Executive
Committee of the USNC/IBP *Recommends* that the SCIBP
formally offer to design and operate a global network to monitor
biologically significant environmental phenomena, to function in
coordination with networks monitoring other environmental
parameters, such as that being recommended by the ICSU *Ad hoc*
Committee on the Human Environment, in cooperation with UN
agencies, participating governments and appropriate international nongovernmental scientific organizations. The above is for
SCIBP consideration at next meeting.' "

In Washington the Federal agencies were becoming concerned with
the international scene as respects environmental quality, and on
August 19, 1969 a meeting of the Federal Council for Science and
Technology Subcommittee on International Aspects of Environmental
Quality was convened. I was invited along with Tom Malone, representing
NAS/NRC, and various others.

Our three-man IBP committee on monitoring was able to meet for the
first time in early October 1969 on occasion of the sixth meeting of SCIBP
in London. A brief interim report was produced for SCIBP to accompany
the three nationally generated papers that we had individually instigated.
Among the principal findings highlighted in this brief report were:

"A global network for monitoring the world's environment in
order to assess short-term and long-term changes is both feasible
and necessary;

"There are various developing networks for environmental
monitoring, which, however, generally neglect the biota in favor
of physical parameters;

"Global monitoring should have as one major aim the acquisition of data pertinent to the global movement of biologically significant materials (such as pesticides, heavy metals, chemical
compounds in general, pathogenic organisms) in the world's air,
waters, and lands and the acquisition of data pertinent to their
effects on the world's ecosystems and their components."

Among the most important recommendations in the report were:

"That ICSU be consulted regarding the possibility that
SCOPE be the international co-ordinating agency for the GNEM;

"That this committee be instructed to develop a plan for global monitoring to be presented at the UN conference in 1972;

"That the committee be instructed to pay particular, but not exclusive, attention to the monitoring of biotic phenomena such as species composition and diversity, fluctuation of population numbers, especially mass increase or decrease in numbers, physiological phenomena such as soil 'respiration intensity,' genetic change in individuals and population, and others of similar kind;

"That the committee consider a framework of *baseline stations* with support from *impact stations* IN AREAS OF IMMEDIATE OR ANTICIPATED MAJOR ENVIRONMENTAL CHANGE and by the use of *indicator* and *sentinel* organisms, including man."

Back in the United States we decided to set up a task force of experts under the aegis of US/IBP to draft a specific plan for global monitoring of the environment. The specific charge to the task force was:

"To clearly determine the environmental parameters to be measured on a global basis to include atmospheric, hydrospheric, lithospheric, and biologic. These decisions must include definitions of spatial density and time resolutions of observations including quality of solar input, and

To prepare a specific modular design for a *primary* monitoring station including interface considerations for data transmission and processing. Modular design is desirable so that 2nd order, 3rd order, 4th order monitoring stations will be compatible with the primary stations."

Glen Hilst, then Executive Vice-President of Travelers Research Corporation, agreed to chair this task force. A highly qualified panel representing a spectrum of anticipated aspects of global monitoring accepted membership on the task force. These included Ed Goldberg (Scripps Institution of Oceanography), Bill Gusey (Senior Wildlife Specialist, Shell Chemical Co.), William T. Houston (Manager, Operations Services, National Lead Co.), Dale Jenkins (Smithsonian), Helmut Lieth (University of North Carolina), Tom Mabry (University of Texas), Bob McCormick (National Air Pollution Control Administration), David Menzel (Woods Hole Oceanographic Institution), Bill Murray (Federal Committee on Pest Control), Ruth Patrick (Academy of Natural Sciences of Philadelphia), George Robinson (Travelers Research Corporation), and Tom Malone (then Senior Vice-President and Director of Research, Travelers Research Corporation).

Under the expert leadership of Glenn Hilst this task force held two working meetings, February 11-12 in Washington and May 14-15 in Boston at the headquarters of the American Meteorological Society. As a result of these efforts I was able to take to the IV General Assembly of IBP at Rome in September 1970, a well done technical report authored by Hilst and his task force entitled, "A Global Network for Environmental Monitoring." This was presented along with a Swedish and a Russian technical report and the summary report of our three-man committee.

Unfortunately, a second version of the Russian technical report entitled, "Global Biological Network Project," and dated August 1970, contained a large infusion of something called heliobiology, the brainchild of an Italian who had seemingly made an amazing impression on the Russian biologists. Professor Piccard, the proponent of heliobiology, made his appearance at the Rome General Assembly to present his arguments for heliobiology in the monitoring program. Happily, this was the beginning and end of "heliobiology" in the IBP and in environmental monitoring.

Our committee report at Rome followed the general lines of the report to SCIBP at the October 1969 meeting in London. However, there were important additional recommendations:

"We recommend to ICSU that:

"It direct the Scientific Committee on Problems of the Environment (SCOPE) to collate the three national reports generated by the IBP *ad hoc* Committee on Environmental Monitoring and further recommend that ICSU present the same to the UN for implementation.

"We recommend to the UN that:

1. The UN accept the plan for a global network for environmental monitoring as submitted by the *ad hoc* Committee of SCIBP through ICSU with the view to its practical implementation.
2. The UN formally organize and assume the responsibility for operation of a global monitoring system, taking cognizance of presently existing capabilities and expanding on these where necessary.
3. The UN draft and adopt a charter covering national and international agreements that are necessary for implementation of a global monitoring system.
4. The UN request that SCOPE serve as a scientific advisory body to the GMS.
5. The SCOPE Commission for Monitoring be asked to undertake detailed planning for the location and coordination of the system, including the problems of central data storage and retrieval.

6.   At an early stage the planning include feasibility studies and method studies at national and transnational levels.

7.   The planning provide for adequate measurement of significant biological parameters even though these may be more difficult to measure than the physical parameters.

8.   Attention be paid not only to the monitoring of the environment in the technologically developed countries but also to the existing problems and potential problems that can occur as the lesser developed countries develop their natural resources."

The report of our IBP monitoring committee was approved by the General Assembly of IBP at Rome. More importantly, it had been approved the previous week by ICSU at its meeting in Madrid. At this same meeting ICSU confirmed the action of its Executive Committee in October 1969 in establishing the Scientific Committee on Problems of the Environment, with a Monitoring Commission as one of its subunits. The task of further developing plans for a Global Network for Environmental Monitoring was referred to this commission, which was in accordance with our committee's recommendation.

Efforts to advance the concept of environmental monitoring received an excellent boost in the public press by Walter Sullivan, science writer for the *New York Times*. Sullivan had been an invited participant in the first U.S. meeting on global monitoring and subsequently had written an excellent story on monitoring which, incidentally, caused something of a flap in the PR section of NAS/NRC. By inviting him to participate in the conference we had made it possible for him to score a "scoop" on the subject. Sullivan went to Rome specifically to cover the IBP General Assembly, and in the October 1, 1970 *Times*, gave the global monitoring efforts an excellent and accurate exposure.

In a smooth transition from IBP to SCOPE sponsorship of the global monitoring effort, the three-man IBP committee became the nucleus of the SCOPE Commission on Monitoring. This suggestion came in a lengthy letter from Tom Malone, writing as Deputy Foreign Secretary NAS/NRC, to Eric Smith, who had accepted chairmanship of SCOPE. Malone wrote:

"Therefore, the presently very exciting and relevant work being pursued under the IBP through its task force on a global network for environmental monitoring (Lundholm, Smirnov, and Blair) is of particular interest to SCOPE. SCOPE should attempt to rationalize the various operational and prospective monitoring systems into a meaningful whole. Indeed, it might be well for SCOPE to invite the IBP-GNEM team to also serve as the nucleus of SCOPE's monitoring effort, at once broadening it to include both additional countries and other program areas such as those cited above."

Others named to the Commission included W. Gallay and Ted Munn of Canada, K. Grasshoff of West Germany, and I. Kunin of U.S.S.R. Lundholm was named chairman.

The work of this Commission received impetus from the fact that on December 10, 1970 Maurice Strong, Secretary-General for the UN Conference on the Human Environment planned for Stockholm in June 1972, had orally requested, and on December 29 formally requested in a letter to Eric Smith that:

"The SCOPE commission on environmental monitoring prepare a report recommending the design, the parameters, and technical organization needed for a coherent global environmental monitoring system making maximum use of available capabilities of existing and planned national, regional, and international networks, together with such data collection and processing centres as may be required."

The first meeting of the SCOPE Monitoring Commission was held in London on January 13-14, 1971 immediately prior to the meeting of SCOPE. The main result of this two-day meeting was a set of terms of reference:

"The functions of the Commission on Monitoring are as follows:
(a) to initiate investigations into the methodology of monitoring, including the selection of suitable parameters, to ensure compatibility of methods and co-ordination of monitoring systems;
(b) to design an integrated, appropriate broad-based monitoring system for air, water, soils, and biota, including man, taking into consideration already existing activities;
(c) to set priorities on parameters to be initially measured commensurate with the urgency of the environmental problems to which they pertain;
(d) to investigate the usefulness of studying past changes in selected parameters in order to establish baseline values and to investigate the possibilities of establishing environmental archives.
(e) A prime component of the system should be a network of background (baseline) stations, far from populations centers, and designed to monitor integrated global values. However, the problems of cities and other areas of major development

should not be neglected, and the commission should consider the development of impact stations and other methods of monitoring those situations where the human impact is critical.

(f) The potentials for future development of special monitoring techniques, such as those of remote sensing, should be carefully considered.

(g) The commission acknowledges and accepts its obligation to design a 'coherent global environmental monitoring system,' as requested of SCOPE by the Secretary-General for the UN Conference on Human Environment — 1972."

The report of the Commission was well received by SCOPE, and the following recommendation was approved:

"The Committee, having considered the possibilities of designing a coherent global monitoring system which would include existing and planned international, regional and national networks, is firmly convinced that such an integrated system is feasible, desirable and deserving of the strong support of the science community, and recommends that the Commission on Global Monitoring take the necessary action, in consultation with the international, regional, and national bodies concerned, to prepare a design for such a system, including the list of variables to be studied, the location of baseline and background stations, and the needs for data collections and analysis centres.

"The Committee recommends that the Secretary take the necessary steps to obtain financial and other support so as to enable the Commission to complete this design prior to the third meeting of SCOPE, so that it may, with the approval of SCOPE, be submitted to the Secretary-General of the U.N. Conference on the Environment shortly thereafter."

Again, the United States took the lead. The Committee for International Environmental Programs (IEPC) had been set up in NAS/NRC in mid-1970 with Tom Malone as chairman. Mainly through the efforts of Malone and of Henry Kellerman, NAS/NRC staffer for the committee, Ford Foundation support for the SCOPE Commission on Monitoring was obtained.

Malone organized a Monitoring Panel within IEPC. In late March 1971 Lundholm and Gordon Goodman of the U.K. came to Washington for strategy planning sessions with this IEPC panel and with others. A working meeting of the SCOPE Monitoring Commission was planned for June 21-24 in Stockholm. Sören Svensson and Gordon Goodman were the Secretariat for the Commission and provided a rough draft of our proposed report prior to the Stockholm meeting. Bob McCormick of the United States participated as a consultant. K. Gupta from Maurice Strong's staff provided an input with respect to the thinking of the staff planners for the U.N. Conference. With the exception of Kunin, all of the members of the SCOPE Monitoring Commission were able to participate, including Smirnov.

Bob McCormick's contribution had a very important impact on the final form of the report. As we began our fourth and final day of work, Lundholm pointed out that the biological variables to be monitored were poorly spelled out in the report. He then asked each person to give a frank opinion of the document. When he asked Bob McCormick, the latter declined as not being a member of the Commission. When pressed, he dropped a small bombshell by saying that he thought the report missed the mark. He rightfully pointed out that the Commission had been asked to design a "Global Monitoring System," but that it had not done so. He felt that the proposed pilot studies were not enough; much more positive action was needed immediately. The discussion of what positive action to take took much of the remainder of the day's session.

We eventually agreed on a concrete program:

(1) For phase one of GEM, there would be a minimum of 10 baseline stations to represent: (a) tundra, (b) northern hemisphere grassland, (c) arctic or antarctic, (d) high mountain, (e) tropical forest, (f) tropical savanna, (g) desert, (h) oceanic island, (i) deciduous forest, (j) coniferous forest. More stations could be approved if desirable. Nomination would be by nations; selecting according to criteria would be by GEMS.

(2) Two of the stations, one tropical and one non-tropical, would be designated as "International Research Baseline Stations." These would be concerned with research of final design of the programs of the baseline stations. International funding and international staffs would be sought for these two stations.

(3) All approved baseline stations would initiate a phase one monitoring program; this would include:
   (a) Monitoring of mercury, lead, polychlorinated biphenyls and organochlorine pesticides in air, water, and soil and in biological organisms representing major trophic levels.
   (b) Monitoring of CO_2 in air, water, and soil.
   (c) Monitoring of particulate matter in air.
   (d) Standard meteorological measurements.
   (e) Determination of the structure of the ecosystem at the respective stations and designation of permanent reference areas representative of this structure.

Our draft report was presented to the First General Assembly of SCOPE in Canberra in early September 1971. A committee chaired by Eric Smith reviewed the report, recommended a few modifications, and the General Assembly then passed a resolution:

"The General Assembly:
Approves the report on Global Environmental Monitoring as amended (to be printed separately) for submission to the Secretary-General of the UN Conference on the Human Environment."

The report was then printed in December 1971 as SCOPE's first publication and formally transmitted on February 6, 1972 to Maurice Strong and the Secretariat for the UN Conference. The key recommendation among 18 in the final report was that, "The United Nations take immediate steps to foster and coordinate among the nations of the world a permanent global environmental monitoring system." With its job done, the Commission on Monitoring was dissolved at the March 16-18, 1972 meeting of SCOPE in London.

How much did the activities originating in the IBP Committee on Monitoring and carried forward by the SCOPE Monitoring Commission influence the development of the United Nations Environmental Program (UNEP) that was developed at and after the Stockholm Conference? This is difficult to answer precisely. However, I believe that these activities were of major significance. The proposed program of UNEP as reflected in sessions of the Governing Council in June 1973, March 1974, and April-May 1975 gives a prominent place to plans for development of a Global Environmental Monitoring System (GEMS) as a part of "Earthwatch." The UNEP secretariat in May 1973 commissioned Ted Munn, representing SCOPE, as a consultant to the Inter-Agency Work Group on Monitoring. His task was to prepare a proposal for an Action Plan for Phase I of GEMS. This was published as SCOPE's third publication later in 1973.

One of the main concerns of the US/IBP was that the IBP could end as an international effort before the necessarily elaborate planning of Integrated Research Programs could be completed or before adequate funding to carry out these programs could be realized. This led me into the second of the three major international efforts in which I became involved in the IBP. In mid-1969, two years of the proposed five-year research phase of the IBP had gone, with most major U.S. programs yet to be funded. Working with Don Farner we had obtained a recommendation by IUBS for the extension of the IBP to mid-1974. At the July 24-25, 1969 meeting of USEC/IBP, a resolution concerning the two-year extension was passed and I was directed to forward it to Jean Baer, President of SCIBP and Barton Worthington, Executive-Director. This was sent by cable under my signature on July 25. It read:

"In view of the fact that the research phase of the International Biological Program is now (July 1969) just getting underway and recognizing that UNESCO has now begun planning a Man and the Biosphere program, the U.S. National Committee for the IBP makes two recommendations for decision of the September SCIBP meeting: 1. The International Biological Program should be continued for a period of not less than two years beyond 1972; 2. Beginning now SCIBP should critically review the International Biological Program and decide which parts (1) should be interrelated with the UNESCO MAB program, (2) should be interdigitated with on-going programs of other UN specialized agencies, (3) should be continued under other auspices, and (4) should be phased out."

The same resolution was sent by letter memorandum from our IBP office in Washington to all members of SCIBP on July 28. Worthington responded promptly, and on July 31, circulated a memorandum to all chairmen of National IBP Committees and National Correspondents, repeating my cablegram and asking for responses.

By the time of the VI SCIBP meeting on October 1-3, 1969, replies had been received from 20 countries. These ran the gamut of support for the U.S. proposal to indication of determination to terminate in mid-1972. Ten were favorable to the two-year extension.

Three of the major participants, France, Hungary, and the U.S.S.R., took no stands on the grounds of inadequate opportunity to consider the matter.

As a result of the divided opinions among the national committees, SCIBP was put in an equivocal position at its October 1969 meeting. However, a resolution was passed and transmitted to V. A. Ambartsumian, President of ICSU, that read in part:

"That in order to take full benefit from results, IBP should continue for a period after 1972, the viewpoint of national committees is divided and a fully coordinated recommendation cannot be expected before the IBP General Assembly of October 1970. However, SCIBP requests ICSU to anticipate providing for not more than two years beyond 1972 for the completion and writing up of IBP projects now in progress."

In a December 5 memorandum to the officers and members of SCIBP, Barton Worthington pointed out that the Executive Committee of ICSU, at its Erevan meeting on October 3, had appointed an *ad hoc* committee, chaired by Harrison Brown, to examine the issue of the two-year extension. The committee was to meet on December 19 in London with Francois Bourliere, Ronald Keay, and Worthington.

The memorandum from Worthington also contained a proposal from Bourliere, President of SCIBP:

> "In the light of present information the President of SCIBP makes the following proposal.
> 1. Phase III of IBP. The programme should continue after July 1972 for a transitional Phase III, lasting about two years. The results of IBP would be assembled and assessed, but this would not be a terminal or negative process. Its purpose would be positive, to bring the full experience and activities of IBP to bear on future programmes in environmental biology and human ecology, and thereby to assist in their detailed planning, in partnership with the new bodies to be created.
> 2. If this proposal were to be accepted by ICSU, and particularly if ICSU wished SCIBP to serve during the transitional period as a non-governmental scientific body in partnership with the Co-ordinating Council of MAB, several matters would need to be examined before a programme and budget for a Phase III of IBP could be prepared. Among them are the following:
>   (i)   the need in the years 1973 and 1974 for meetings of SCIBP, its Bureau and possibly the Assembly;
>   (ii)   the need by the sections and working groups of SCIBP for meetings, consultations and related activities;
>   (iii) depending on estimates of the work-load which would be entailed, the degree to which IBP's central and sectional officers should be maintained."

The *ad hoc* committee met as scheduled and, with the SCIBP members who were present concurring, recommended that there should be a two-year continuation of IBP to June 30, 1974.

The SCIBP Bureau at its Amsterdam meeting in early April 1970 had an extended discussion of the urgent need for an orderly transition from IBP to whatever its successor might be and passed a resolution to the effect that an extension was urgent and requesting the Secretariat to poll the national committees on the subject of supporting the recommendation to ICSU by the biological science unions that the IBP be extended to 1974.

At a meeting of USEC/IBP on June 18-19, 1970 Harve Carlson expressed NSF views toward extension of the IBP:

> "• NSF expects the types of multidisciplinary research as represented by US/IBP-IRP's to be continued indefinitely into the future.
> "• NSF has not yet established a formal policy supporting extension of IBP to 1974. Informally, NSF people are in favor of extension."

The important General Assembly of ICSU at Madrid in September 1970 approved the extension of IBP for two years, now to terminate at the end of June 1974. In the United States the IBP biome research was destined to continue for at least two years beyond this date, and the Analysis of Ecosystems Program in NSF was to become a seemingly permanent entity there.

The losing battle on the international scene in which I was deeply involved was the one to try to ensure a successor for SCIBP as an umbrella for international ecosystem science at the non-governmental level. The period of 1969-1972 saw the emergence of several potentially competing initiatives with respect to international environmental programs. No planning could ignore the possible significance of programs that might emerge from the coming June 1972 UN Conference on the Human Environment. UNESCO's Man and the Biosphere (MAB) was being touted as the successor to IBP, with Michel Batisse promoting this idea at every opportunity. In this he had strong support from Barton Worthington and some others in the IBP organization. SCOPE became a potential candidate for successor to SCIBP after its establishment in September 1970. Senator Magnuson's S. B. 399 to establish an International Center for The Environment (ICE) was mentioned earlier. Tom Malone was pushing ICE in SCOPE and ICSU.

USNC/IBP gave serious thought to the future of international ecosystem science beyond termination of the IBP. My telegram of July 25, 1969, conveying USNC/IBP's recommendation of a two-year extension also carried the recommendation:

"Beginning now SCIBP should critically review the International Biological Program and decide which parts (1) should be interrelated with the UNESCO MAB program, (2) should be interdigitated with on-going programs of other U.N. specialized agencies, (3) should be continued under other auspices, and (4) should be phased out."

The resolution passed by SCIBP at its London meeting of October 1-3, 1969 in response to this recommendation urged in part:

"That ICSU should take into account the importance of continuation of the appropriate IBP-generated research under sponsorship of either intergovernmental or non-governmental bodies, or both."

In a letter to Dick Oliver dated December 5, 1969, Hugh Southon, SCIBP Executive Secretary, discussed the situation as regards the phase out of IBP:

"It seems clear already that the pattern of the transition period will be unclear for some time and that much work will be required to define the most efficient way of developing international activities in environmental biology on the two levels that are required at present. While an inter-governmental programme may well focus more sharply on managed ecosystems and situations than is IBP, it is essential at the same time to foster the fundamental studies on which the methodology of the wider programme will rely heavily and which are themselves in urgent need of considerable extension.

"This development calls for a strong non-governmental body under the ICSU umbrella, not only to be advisory to any co-ordinating body set up by UNESCO and the other UN agencies, but to participate actively in the detailed planning, and where appropriate, in the implementation of the programmes."

In the United States we determined to take the lead in establishing a plan for the phase-over of IBP activities at such time as the IBP would end. On June 11, 1970 I circulated to USNC/IBP a proposed "Phase-Over of IBP to Other Programs." Pertinent to the continuation of ecosystem studies at an international level, this document read:

"IRP's — Expand to World Program of Biome Studies (delete IBP Sections and Themes).
  A. ICSU — SCOPE will provide umbrella for research.
  B. Delete Distinction of Environmental & H.A. Components.
  C. SCOPE Commission on "Man's Interactions with Earth's Ecosystems" (MIWEE) to be International Governing Body (replaces SCIBP).
  D. National Committees Under National Academies (to replace IBP National Committees).
  E. Implementation
      (1) (1971) Formation of *Ad hoc* committee for MIWEE and meeting of same.
      (2) (1971-72) Transformation and restructure of IBP National Committees into MIWEE Committees.
      (3) (1973) Transition of IBP sponsored research to MIWEE sponsorship.
      (4) (1974) Preparation of reports summarizing and synthesizing IBP research.
      (5) (1972-74) Regional and International Biome Conferences to plan continuation and expansion of internationally coordinated research.

"International Biome Research

Basic objectives: Through multidisciplinary and internationally coordinated and cooperative studies of the world's major ecosystems, to gain an understanding of the *structure* and *function* (especially productivity) of these systems, and to gain an understanding of man's interactions in, and with other components of, these systems."

At a June 19, 1970 session of IEPC in Washington, I presented this same proposal. During this presentation I pointed out that:

"The IBP research could either go to UNESCO, as part of its MAB (Man and Biosphere) program or to SCOPE. SCOPE would appear to offer the preferable alternative. The question yet to be resolved was, of course, the financial support for the type of research involved in the study of the selected ecosystems."

In an extended discussion, Henry Kellerman, in response to a question regarding MAB's potential as a successor to IBP, reported that discussions with the Director General and other key officials of UNESCO had demonstrated that, while UNESCO appeared eager to absorb IBP at an early date, it had not provided adequate financial or organizational arrangements for a take-over. In Fred Smith's view expressed during the morning session, UNESCO's terms of reference, moreover, appeared to be too narrow, leaving no room for "action."

Abe Chayes expressed considerable doubt about UNESCO's capacity to sponsor the type of program that was under consideration.

On July 2, 1970 Tom Malone and I jointly chaired a small meeting of US/IBP and IEPC persons to discuss the program and structure of SCOPE. The meeting had access to the report of ICSU's *ad hoc* Committee on Problems of the Human Environment and my paper on "Phase-Over of IBP to Other Programs." Development of an ICE was a major point of discussion.

On September 9, 1970 Barton Worthington, Scientific Director of IBP, sent to Eric Smith a memorandum entitled "A Note by SCIBP for Consideration by SCOPE." In this he raised five questions. Two of these and the answers provided on September 24 by Eric Smith are pertinent here:

"Does SCOPE as a 'special committee' of ICSU contemplate a global programme of co-ordinated projects such as SCIBP was set up to run? Or does it envisage operating in relation to other programmes more on the lines of SCOR and SCAR?

"SCOPE sees itself as a body which, at least in its early stages, will seek through the discussion of its main committee and by the more easily attained member communication and more frequent meetings of its Bureau, to encompass in its thinking the entire range of environmental problems of the atmosphere, hydrosphere and geosphere. It will, however, be rigorously selective of the tasks it will undertake and modest in the rate of expanding them so as not to overreach its capability until such time as it is better organized and able to undertake a wider responsibility. SCOPE will look initially to other scientific organizations and agencies to undertake research which its investigations show to be necessary, but if none is able or willing to include this research in its programme, SCOPE will need to consider how to promote the work within its own organization.

"If SCOPE envisages its own world programme, is it likely that it would be in a position in the biological sciences to develop some continuing activities of the IBP?

"Although SCOPE does not envisage its own world programme developed in the IBP manner it would welcome the opportunity of developing activities which IBP may not be able to complete within its lifetime and to that end would wish to receive the advise and cooperation and perhaps incorporation of IBP in planning its present and future activities."

In the United States, one significant influence was the publication of a report entitled, "Protecting the World Environment in the Light of Population Increase," by the Office of Science and Technology (OST) in the Executive Office of the President. Published in December 1970, this document included in its recommendations: "To continue the International Biological Program and to select those elements which have utility beyond the expiration of the Program."

At the meeting of USNC/IBP on December 9-10, 1970 the possible role of SCOPE as a follow on to SCIBP was discussed:

"• SCOPE is very aware of IBP and of the systems analyses that IBP has initiated. SCOPE would like to see these studies continued.

"• SCOPE, at this time, does not see itself inheriting the sponsorship of IBP studies. SCOPE intends to design its own program.

"• SCOPE needs advice on the kind of body that should continue sponsorship of selected IBP initiated research. — Implicit here is that SCOPE will help cause such a body to be created — perhaps outside SCOPE's area of interest — perhaps within an expanded view of SCOPE's purview."

At the meeting of SCOPE in January 1971, Tom Malone presented the case for the establishment of an ICE. Among the several functions proposed for ICE was to:

"Provide coherence and integrity to continuation of research on biomes, initiated under the auspices of IBP, including integration of the study of human adaptability, interactions of components of representative biosystems, methods of increasing biological productivity and biome modelling."

At this same meeting I mentioned the fact that USEC/IBP had had discussions with Maurice Strong with respect to the relevance of IBP biome studies to the UN Conference agenda. Several aspects of the IBP research effort had a distinct bearing on the broader problems considered by the UN, namely the factor of predictive capability, especially predictive economic capability, and the relationship to an organized monitoring system. He wondered how SCOPE was planning to handle, or relate its efforts to, the continuation of global biome studies. I received no definitive answers.

The next step in the attempted development of a successor to SCIBP came at the meeting of the SCIBP Bureau (Executive Committee) at Budapest on June 3, 1971. At Budapest the Bureau adopted a resolution for ICSU officers, also meeting at Budapest, but aimed at SCOPE. This read:

"The SCIBP Bureau: draws the attention of the officers of the ICSU to the following informal resolution, agreed on 3 June 1971.
Recognizing that the life of the IBP and its management structure will end in 1974;
Calling attention to the demonstrated success of the IBP as a non-governmental coalition of scientists for attacking global problems;
Observing the need for continued attention to the major biological problems of the type to which IBP has been addressed for many years in the future;
Emphasizing the consequent need for a strong mechanism for continuation of the contribution of the world community of scientists to the solution of the problems;
"Recommends that ICSU consider its responsibilities and opportunities, including possible institutional arrangements, for the continuation and development of research in environmental biology at the international level including appropriate elements of the IBP."

This was accepted by the ICSU officers on June 5 and referred to SCOPE. At Budapest an *ad hoc* committee of Jim Cragg, Otto Frankel, Joe Weiner, Barton Worthington, and myself was appointed to make recommendations with respect to the resolution. I agreed to produce a first draft of a memorandum to SCOPE on the subject. I wrote most of this on the long flight home from Budapest and on June 11 circulated the draft to the other members of the *ad hoc* committee and to Francois Bourliere for comment.

The document produced by our *ad hoc* committee attempted to explain the intent of the SCIBP resolution:

"In sum we wish to speak to:
(1) the nature and potentialities of the internationally co-ordinated research that has had its genesis under the IBP,
(2) the advantages of the kind of international co-ordination and co-operation among environmental scientists at the non-governmental level that has developed under IBP, and
(3) the urgent need for institutional arrangements at the non-governmental level that will provide for continuation of this co-operation on a global scale, and will assure co-operation between the non-govermental and inter-governmental sectors of research in environmental biology."

It made as strong an argument as we felt possible for a continuation of an international management structure at the non-governmental level:

"The biome approach, with man as an integral component of the biome model, and with interbiome synthesis and modelling of major ecological processes, promises to provide increasingly valuable predictive models for utilisation and management of the world's resources. The scientifically based management techniques which will emerge from these studies will provide the knowledge necessary for the maintenance and improvement of the quality of man's environment.

"Studies of this type will be in peak operation in some partici-pating countries at the official termination of the IBP in 1974, and the task will be far from complete. Comparable studies on the important tropical ecosystems may be barely started or still in the planning stage. Therefore, we feel that it is of the highest im-portance that a mechanism be provided for continuation and ex-pansion (especially in the less-developed countries) of the biome related studies at non-governmental level. ICSU is the only existing non-governmental organization with the strength to provide this mechanism. SCOPE, as the environmentally-oriented dependency of ICSU, would seem the logical home for this co-ordinating structure.

"It must be made very clear that we are not proposing an extension of IBP and of SCIBP beyond 1974. What we are suggesting is a committee that might carry some such label as 'Committee for Assessment of World Ecosystems.' It might have no more than two subsections: (1) Biome Structure and Function (with man and his effects included here), (2) Interbiome Process Studies.

"While recognizing that machinery for stimulating and co-ordinating national activity in these fields at governmental level is to be established by the Man and Biosphere programme (MAB), we feel that there are many activities which can be undertaken more effectively at non-governmental level, and it is in respect of these that the proposed committee of ICSU is needed. The role of this Committee in organising and coordinating world research would be similar to the role of SCIBP but with respect to a restricted field of study. The ecosystem studies which it fostered would interact with those of MAB to mutual advantage. Indeed the Committee would have an important function in advising MAB, perhaps becoming the non-governmental counterpart to MAB ready to organise and co-ordinate international activities which can most effectively be undertaken non-governmentally.

"In conclusion it must be emphasised again that the biome studies to which special attention is devoted in this memorandum are only a part of the IBP, but they need to be considered in a very wide context which includes the pressures on the environment from human population increase, and all forms of technological activity, including pollution. Special attention should be given to those ecosystems which are among the most vulnerable and the least known, namely those of the tropics, sub-tropics and sub-arctic."

On June 25 Barton Worthington circulated our memorandum to the SCIBP Bureau, Sectional Convenors and Scientific Coordinators for comment prior to submitting it to SCOPE on July 15 for consideration at the SCOPE General Assembly on August 31-September 4, 1971.

In an April 30 letter to me, Ronald Keay had transmitted a paper which he had prepared for the British National Committee for the IBP. One pertinent comment was:

"It must be emphasised however that MAB is an inter-governmental programme and it cannot be expected to provide the same kind of framework for international scholarly discussion and co-operation which IBP has provided. For instance, in convening meetings of experts on particular subjects, UNESCO will have to pay attention to political issues (e.g., balance between different political areas, exclusion of South Africa etc., etc.) and the selected experts will have to be approved by their governments."

At the July 15-16, 1971 meeting of USEC/IBP and Component Coordinators, we discussed the future of ecosystem analysis beyond the life of the IBP. Concerning MAB, Bill Milstead:

"...expressed his unhappiness, and that of others, with MAB because it is an inter-governmental arrangement. He feels that the Biomes should be affiliated with MAB rather than being put under MAB, since it might be difficult to get funding through NSF or any other Federal agency if the Biomes are part of an inter-governmental organization."

The SCOPE General Assembly at Canberra, on September 4, 1971, adopted the following resolution:

"The General Assembly:
recommends that a working group be established to examine the ways in which SCIBP Programs on Biomes, production processes, and human adaptability might best be developed within SCOPE and in relation to MAB and to report to the 4th meeting of the Committee."

The Working Group recommended at Canberra was soon set up with Francois Bourliere as Chairman. Of those invited to participate, A. R. Clapham of the U.K., Joe Weiner, Gilbert White of the United States, and I, agreed and participated in an Amsterdam meeting on January 4, 1972, immediately preceding a meeting there of the SCOPE Bureau. This task force recommended that SCOPE adopt a Program for Analysis of World Ecosystems (PAWE), with initial emphasis on tropical ecosystems. Meeting on January 6, the SCOPE Bureau agreed with our group and passed a resolution that read in part:

"Having considered the report of the SCOPE/SCIBP Working Group on ways by which SCIBP programmes on Biomes, Processes and Human Adaptability might best be developed within SCOPE and in relation to MAB, the Bureau agreed:
1. to adopt a Program for Analysis of World Ecosystems (PAWE), as shown in outline in the annexed diagram, as a means of extending the conceptual and methodological approaches to ecosystem analysis and modelling developed for

IBP through its several sections by taking into account additionally socio-cultural variables. Initially, PAWE should concentrate on humid tropical ecosystems. This would include a major contribution by SCOPE to the preparations for MAB, and SCOPE would offer to provide the conceptual framework and necessary methodologies for MAB Project I (Ecological Effects of Increasing Human Activities on Tropical and Subtropical Forest Ecosystems).

2. to develop at a later stage other elements in the Program for the Analysis of World Ecosystems, such as grazing lands and forests, incorporating the results of the IBP projects. SCOPE would assume that SCOR would take the initiative in developing and advising on a program for marine ecosystems."

I was designated by the Bureau to organize a task force to prepare a plan for implementation of this program and report back to a London meeting of SCOPE in March. I had a draft of the plan for activation of PAWE completed by mid-February, and on February 25, Tom Malone mailed copies to members of SCOPE. The objectives of PAWE as outlined in this document were:

"1. Provide an international management structure for research on and predictive modelling of the world's major ecosystems, with man and his activities as important components of the system, and promote the development of national arrangements for participation in these efforts.

2. Coordinate conceptual and methodological approaches by national and regional research teams taking into account socio-cultural variables.

3. Organize and sponsor interregional and interbiome conferences of process specialists.

4. Promote and develop regional projects where these are urgently needed to fill gaps in our knowledge of the functioning of the world's ecosystems.

5. Strive to achieve the highest possible level of basic scientific understanding of the structure and function of the world's ecosystems and of the social implication of man-induced changes and to make this information, including predictive models, available as a basis for man's most wise use of, and interaction with, world ecosystems.

6. Have as a very long range objective the capability of producing models of those ecological processes that pertain to the entire biosphere.

7. Provide on request advice to international agencies (e.g. UN and its specialized agencies; IUCN; etc.) through SCOPE and/or ICSU. Establish linkages at all stages with decision makers with respect to environmental matters."

The recommendations for immediate action were:

"It seems very urgent that some immediate actions be taken in order to maintain the momentum in biome research that has been achieved in the IBP and in order to maintain and extend the present level of international cooperation in biome research. The main actions needed are:

1. The establishment of a SCOPE Commission for the Program on Analysis of World Ecosystems, as recommended by the SCOPE/SCIBP Work Group and approved by the SCOPE Bureau at Amsterdam, with representation from all geographic regions and scientific disciplines relevant to the subject area under study.

2. The establishment under the Commission of a Task Force on Tropical Ecosystems, to concern itself initially with the humid tropical forest and tropical savanna ecosystems with representation from all geographic regions and scientific disciplines relevant to the subject area under study.

3. Designation of a secretariat for PAWE, in SCOPE Headquarters, and acquisition of funding to support same (estimated $30,000 annually).

4. Discussions with SCIBP and its appropriate task forces concerning the possible phase-over to an appropriate task force or task forces of the SCOPE Commission for PAWE, before IBP formally terminates in mid-1974 for the international coordination of selected activities, such as the biome studies of grazing lands and of temperate-forest lands.

5. Preparation of a proposal to funding sources for support of the initial activities of the Task Force on Tropical Ecosystems."

I went to the SCOPE meeting of March 17-18 in London with considerable input and strong support from USNC/IBP and from IEPC. I had lists of suggested nominees for both a PAWE Commission and for a Tropical Task Force. After extended, largely favorable, discussion, SCOPE gave its endorsement to the PAWE plan in a resolution that read:

"*Agreed* that a commission for the Programme for the Analysis of World Ecosystems (PAWE) be established with the terms of reference given below 1-5 and with the provisos:

(i)   that the commission should report back to the committee of SCOPE early in 1973 on the development of the programme in relation to MAB,

(ii)   that the initial membership should include Prof. W. F. Blair (Chairman) and Dr. R. O. Slatyer, with other members to be appointed in consultation with the bureau, and an observer representing the MAB programme of UNESCO, and

(iii)   that financial provision from SCOPE funds should be made for a meeting of the commission during 1972.

"TERMS OF REFERENCE

1.   To include within its membership representation from geographic regions and scientific disciplines relevant to the subject area under study.

2.   To establish a Task Force on Tropical Ecosystems that would concern itself initially with the humid tropical forest and tropical savanna ecosystems with representation from geographic regions and scientific disciplines relevant to the subject area under study.

3.   To designate a secretariat for PAWE, in SCOPE headquarters, and acquire funding to support same (established $30,000 annually).

4.   To discuss with SCIBP and its appropriate task forces the possible phase over to an appropriate task force or task forces of the SCOPE Commission for PAWE, before IBP formally terminates in mid-1974, for the international coordination of selected activities, such as the biome studies of grazing lands and of temperate-forest lands.

5.   To prepare a proposal to funding sources for support of the initial activities of the Task Force on Tropical Ecosystems."

The next major event in this scenario was the UN Conference on the Human Environment in Stockholm in early June 1972. Representatives of ICSU were invited to participate as observers. Along with representatives of IUCN (International Union for the Conservation of Nature), we were given the only office space provided for non-governmental groups at the conference. We had badges which admitted us to the observer's balcony as well as 4 badges, which we interchanged among us, that admitted us to the floor of the conference hall. This show of concern by Maurice Strong for the non-governmental body of science represented by ICSU seemed to annoy the bureaucrats of the UN agencies and I am sure that it contributed to Michael Batisse's subsequent efforts to destroy PAWE as a potential competitor with MAB for funding by UNEP.

During the course of the Stockholm Conference, we attempted to compromise with Batisse and his MAB program. In a meeting of June 15, between ICSU, IUCN, and UNESCO representatives, I asked Batisse if he saw any reason why MAB could not support any activity that was a component of a biome study (U.S. style). He said no. My notes of the meeting contain the statement, "If we can hold him to this there should be no problem of cooperation and coordination."

In a paper produced subsequent to the Stockholm Conference the UNEP Secretariat identified PAWE as a possible source of further information on analysis of ecosystems. This, in my view, doomed PAWE to persecution by Batisse until SCOPE/PAWE could be eliminated as a potential competitor for UNEP funds in the environmental area.

The fifth and final General Assembly of the IBP was held in Seattle in early September 1972. This meeting had been scheduled originally for Paris in the expectation of Worthington and others that the IBP would simply pass to UNESCO and MAB, and thus a final meeting at UNESCO headquarters would be appropriate. My arguments that no meeting of SCIBP, or the SCIBP Bureau, or of an IBP General Assembly had been held on this side of the Atlantic prevailed and Seattle was chosen as the site. During the General Assembly I chaired an informal session to discuss PAWE, the participants having been provided with my June 19 version of the program. Denouncement of IBP failures and of the proposed PAWE by such representatives of LDC's as Osvaldo Boelcke of Argentina, B. R. Seshachar of India and others convinced me that the opposition had been thoroughly programmed before the session began.

In spite of the discouraging response in the Seattle meeting, we moved ahead with plans for implementation of PAWE and with good support in SCOPE. As a next step toward realization of PAWE, a small workshop was planned for Helsinki to precede the ICSU General Assembly in mid-September. The terms of reference for this workshop were:

"1. To examine in detail the terms of reference for the PAWE Commission agreed by 4 SCOPE.
2. To consider appropriate means for continuing consultation with MAB by which to ensure that the best use is made of the experience of IBP, that unnecessary duplication and overlapping of effort between MAB and PAWE is avoided, and that the complementary elements in both programmes are brought together and harmonized.
3. To consider membership of the commission.
4. To formulate a proposal for the funding of selected elements of PAWE with view to a more extensive proposal later."

I chaired this work group; additional participants included Paul Baker and Dave Reichle of the United States, Francois Bourliere, R. Misra of India, Bengt Lundholm, Joe Weiner, and Hugh Southon, Executive Secretary of SCOPE. This group at Helsinki developed a two-phase plan for PAWE, with a proposed budget of slightly over $200,000 for Phase I. It also identified potential members for the PAWE commission, with a wide geographic and disciplinary representation.

At Helsinki one half-day symposium had been organized by Tom Malone under the title of "Environment 1980." Peter Thacher of Maurice Strong's UNEP staff was the first speaker after opening remarks by Frans Stafleu, Secretary General of ICSU. Thacher spoke strongly of the role of the world's scientists:

> "I believe that the activities of the scientific community, and particularly certain specific activities by this Assembly of ICSU, have led states at the political level to a better appreciation of their interdependence and the need to avoid actions which are mutually harmful. The Stockholm Conference made it possible for states to set in motion a number of activities which call for contributions from the scientific community which, for full effectiveness, must be voiced authoritatively at the international level.
>
> "I suggest that ICSU contains within its ranks a growing capacity to predict the outcome within the biosphere of many of man's actions, that ICSU has itself shown in practice how to set up the means of advance multidisciplinary assessment, and that the decision-makers who have authority over activities which give rise to concern are demonstrably ready to receive and take account of objective, authoritative assessments at the international level.
>
> "The opportunity for the scientific community to contribute to wiser, more rational choices on the part of decision-makers has never been greater. The willingness of governments to act has been affirmed in their decisions at Stockholm; they approved an Action Plan by which to proceed and are committed to provide the means. We look to you, scientists assembled in ICSU, and to your colleagues, to continue to contribute your skills and knowledge, as a part of Earthwatch to provide the assessments on which more rational decisions will be made to protect and enhance the human environment."

I presented the plans for PAWE. Other speakers were Lee Talbot of the United States representing IUCN, A. Buzzati-Traverso, Assistant Director General for Science of UNESCO, and Warren Wooster of the United States as President of SCOR.

The next major crossroads for PAWE was the meeting of the SCOPE Bureau in Ottowa on October 8-11 where I again presented the PAWE program. Some modifications were proposed, the principal one of which was Eric Smith's suggestion that PAWE be expanded to include marine ecosystems. This modification was later used against us by Batisse, who argued that PAWE was trying to cover the entire globe, while MAB was sharply focused. Following the Ottowa meeting, I incorporated the recommendations approved there into the PAWE plan and circulated the same to the members of the PAWE task force on October 18.

In the United States, support of PAWE was strong in late 1972. Tom Malone, as Secretary General of SCOPE, convened a December 13 meeting in Washington, "to critically examine the ... PAWE document and to do planning for the next several months of the PAWE undertaking."

Fred Smith, who was one of the participants in the December 13 conference, two days later wrote to Henry Kellermann, NAS/NRC Staff Officer for IEPC:

"1.   PAWE, as a mechanism for international communication among research programs on ecosystem analysis, and as a means of focusing attention on needs to develop new programs, provides an extremely important function that is exactly what ICSU should do. These sessions that IBP has been holding (internationally) have been mind-stretching to an extraordinary degree, and a lot more of the same remains to be accomplished. In fact, this process never ends.

2.   SCOPE must establish COPAWE as a first step, since that commission, willy nilly, must be responsible for the development of PAWE. A dozen good men is far more important than details of their scientific fields — but they should have competence related strongly to the objectives of PAWE.

3.   COPAWE should find a Program Director designate, who would be the 'principal director' of a proposal for funding. The prospectus we have been kicking around could be approved as a guideline by SCOPE, transmitted to COPAWE, and then to the designate. He and COPAWE would have to flesh it out — using a small study group or any other way that seems feasible.

4.   Alternatively, it may happen that PAWE will be so institutionalized in everyone's plans (Strong, etc.) that only a solid and respectable COPAWE, plus an obviously qualified designate-director, plus a short prospectus, would be needed in a proposal for funding. I think this should certainly be tried, both to see if it works, and as a means of early contact with prospective funders.

5. PAWE must be extremely selective with respect to the programs for which it will offer international com- munication. High relevance to perceived needs, and good opportunities for improvement of scientific quality and/or productivity, should be major guidelines."

On December 21 I transmitted to Hugh Southon of the SCOPE Secretariat the latest, and as it proved to be, the final, version of PAWE along with recommended names for appointment to PAWE and its task forces.

The final axing of PAWE came in a "Summit" meeting between ICSU and UNESCO at UNESCO House in Paris in early January 1973 on occasion of the fifth meeting of SCOPE. Batisse's biopolitics were never in better form, and the ICSU group knuckled under completely. The out- come was essentially that SCOPE activities under its PAWE must be subservient to MAB.

I returned home, and on January 18 wrote to Eric Smith resigning my chairmanship of the now nominally existent SCOPE Commission of Predictive Analysis of World Ecosystems. My rather bitter letter read in part:

"The basic reason is, of course, the overt and covert op- position to PAWE that has existed in UNESCO since the inception of this activity and the inability of unwillingness of SCOPE to move this program ahead in the face of this opposition. The campaign to destroy PAWE reflects no credit on the UNESCO personnel who have promoted it. The vulnerability of SCOPE to this campaign is most discouraging.

"Although I remain convinced of the potentially great contri- bution of the PAWE concept, especially in relation to the developing UN Environmental Program, I have reached a point of complete frustration, hence this decision."

PAWE was dead in spite of our repeated efforts to compromise with Batisse and his MAB program.

5

The Operational Phase of US/IBP

I will deal rather briefly with the operational, as opposed to the preparatory and promotional, phase of the IBP in the United States. A very early landmark in the U.S. program was the May 1967 grant of $71,400 by NSF for planning of the Analysis of Ecosystems (AOE) program, with Fred Smith as Program Director. In June of 1968 this was augmented by a two-year grant of $224,300 for central management of AOE. On recommendation from AOE, the first funding for a biome study was provided in 1968 with a grant of $350,000 from NSF and $50,000 from AEC to initiate research in the Grasslands Biome, headed by George Van Dyne of Colorado State University. This was increased to $851,000 later in 1968 and to $1.8 million in 1969.

Four additional biome studies under AOE were funded at operational levels before the end of 1970. The Eastern Deciduous Forest Biome received funding of $186,400 from NSF and $40,000 from AEC in late 1969 and saw this increased to $1.2 million by September 1970. The Desert Biome received a grant of $20,200 from NSF in January 1970; it later received an operational grant of $654,000.

The Coniferous Forest Biome received a six-months planning grant from NSF of only $3,200 in June 1969, but in September 1970, it had an operational grant of $450,000.

The Tundra Biome had seemed an unlikely candidate for support, but the oil strike on the North Slope of Alaska changed all of that. By May 1970 the Tundra Biome had support in the order of $450,000 from NSF and a combination of sources, including private industry.

Other IRP's other than AOE were receiving support before the end of 1970. The Hawaii Subprogram under Origin and Structure of Ecosystems in mid-1970 received $229,000 of support from NSF. The Upwelling Ecosystems program headed by Richard Dugdale received $161,500 from NSF in February 1970. Six components of the Origin and Structure of Ecosystems received NSF grants totaling $358,500. Four components of the Biological Control program received a total of $146,100 in NSF support in late 1969 and early 1970.

The Conservation of Ecosystems Program, headed by George Sprugel, received a grant of $52,600 in July of 1970 under auspices of AIBS.

149

The Human Adaptability components of US/IBP were having greater difficulty in obtaining support than the environmental components apparently because of the greater inflexibility of NIH, their logical supporter, than of NSF. However, Jim Neel's study of South American Indians did receive NSF support in the amount of $62,500 on April 1, 1970 to augment his continuing support from AEC. Also, a two-year grant from NSF was provided in the amount of $90,000 to support Bill Laughlin's IRP to study the Aleuts.

One managerial crisis faced by the USNC/IBP was the resignation of Fred Smith as Program Director of the Analysis of Ecosystems. As initially established, AOE was to have seven components, of which six would be biome studies, and the seventh would be a "Central Program." The role of the Central Program was intended to be one of service to the separate biomes and, additionally in Smith's words, to do research on "generalized analysis of principles applicable to all biomes." In a progress report on AOE circulated in March 1969, it was estimated that support was needed for the total AOE in the amount of $15 million annually. This report also carried the statement that: "Unless several biome studies are fully supported, the continued support of the Central Program is not recommended."

In June 1970 Smith, who had moved from Michigan to Harvard University, resigned as Program Director of AOE, thus throwing the Central Program into confusion. In his memorandum of resignation he expressed frustration at prospects of inadequate funding for the program. He recommended transfer of the management activities of the Central Program of AOE to USNC/IBP and abandonment of the Central Program.

John Kadlec, who had been Program Coordinator for AOE, having been named that in late 1967, remained at the University of Michigan and continued the activities of his office pending a decision by USNC/IBP as to what course to take. At its June 1970 meeting USEC/IBP regretfully accepted Fred Smith's resignation and made plans for continued coordination of the biome studies. It was decided that the biome directors as a group would act in this capacity, and that an office for coordination of all of the Integrated Research Programs in the Environmental Component of US/IBP would be established at the University of Texas at Austin. A similar coordinating office for the Human Adaptability Component was to be set up at Penn State University. This arrangement was approved by NSF, which provided financial support for the two offices. The IBP office at NAS/NRC was affected to some extent by the establishment of the two offices. However, it continued to be the main point of contact with Washington agencies and activities, including NSF as main sponsor of IBP in the United States. It continued to be the main point of contact with SCIBP and the international aspects of IBP. It continued to support the activities of USNC/IBP and USEC/IBP. The two component offices could give their attention to the IBP research itself, including the arrangements for technical meetings of the IBP researchers. The coordinators of these two offices were also responsible for drafting annual reports No. 5 and No. 6 of US/IBP. The Austin office also produced the IBP Interamerican News until its unfortunate demise in 1971.

The Austin office was opened in January 1971 with Bill Milstead as Coordinator and Norris Loeffler as assistant and Public Information Specialist. These served a two-year term. Orie Loucks became coordinator for a year after Milstead returned to his university. The H.A. office was opened in May 1971 with Frank Johnston as coordinator and Sharon Friedman as Public Information Specialist. Subsequent coordinators were Joel Hanna and later Mike Little. When John Reed replaced me as Chairman USNC/IBP in mid-1972 I continued as an *ex-officio* member of the committee, and I continued to supervise the Austin office.

As the biome directors looked toward the date in mid-1974 when the IBP apparatus would no longer exist, they were concerning themselves with the future of ecosystem research after IBP. Bill Milstead, who in his capacity as EM Coordinator had visited and talked with all biome directors, told the USEC/IBP meeting of July 15-16, 1971 of the suggestion that the biomes approach the new Institute of Ecology in a block to negotiate terms of affiliation. A planning meeting of biome directors was held in September 1971, and another was held on occasion of the meeting of USNC/IBP at Oak Ridge in early November. At that time the plan was for agreement on a 10-year program of cooperation under an AOE program.

A series of meetings, back to back, in mid-January 1972, was crucial to the phase-over from IBP to TIE in ecosystem research. The first of these was a meeting of the biome directors on January 18 and the morning of January 19. At this meeting it was agreed to meet with Art Hasler, TIE Director and Bob Inger, Chairman of the Board of Trustees of TIE, to discuss possible affiliation of the biomes with TIE. This meeting took place on the afternoon of January 19. The general conclusions that were reached are summarized in the minutes of the meeting:

"a) TIE does not intend to compete with the IBP Biome programs sponsored by the USNC/IBP. — Rather, TIE intends to pick up on these efforts where IBP leaves off.

  b) TIE does desire to explore the human component in context with the transitional phase of selected IBP programs to eventual sponsorship by TIE.

  c) A joint Task Force between USNC/IBP and TIE should be considered by both activities — to establish a mechanism for selecting those U.S. programs and plans within IBP that TIE should sponsor."

The USNC/IBP met on January 20 and spent much of the time on discussion of how to bring the HA and EM components of the IBP into closer coordination. The slogan was used, "to put man into the ecosystem." A large meeting of USNC/IBP with representatives of Federal Agencies, including members of the Interagency Coordinating Committee, took place on January 21. Included in the purpose of this last meeting was to consider what follow-up programs to IBP should be considered. Among the generalizations from this meeting reflected in the minutes was:

> "The Institute of Ecology (TIE) may be the most logical institution to evaluate "New Directions"—and TIE may be the most desirable institution to assure the future integration of HA-Biome research programs. The follow-up program or programs to IBP, whether integrated by TIE or by some other activity, should carefully and explicitly include the mission interests of *all* concerned agencies."

USEC/IBP met the same day and approved the establishment of a joint IBP/TIE Task Force. On January 25 I wrote to Art Hasler naming the US/IBP members of the Task Force: Fred Wagner (Chairman), Dick Dugdale, and George Van Dyne. My letter to Hasler included an expression of the sentiments and suggestions of USEC/IBP:

> "Our IBP Executive Committee expressed a desire that the Joint Task Force should work in the context of "Man and the Environment"—thus rectifying a dichotomy that has existed through much of the life of IBP. We feel that only recently has our USNC/IBP made significant progress on this point, but that our recent progress indeed has been significant.
> "Our Executive Committee also strongly suggested that TIE might establish a type of executive directorate within an appropriate place in the TIE structure that would assure coordination and planning among and across all Biomes. Personally, I am quite confident that the Task Force will explore this and the many other needs involved in this transitional venture to the full satisfaction of TIE and the USNC/IBP."

However, this task force never functioned except in informal discussions among the IBP appointees. At the USNC/IBP meeting of November 27-29, it was agreed that a new Task Force would be established and this was appointed by John Reed on January 11, 1973 with myself, Dave Goodall, Lee Kline, Terry Rogers, and Kenneth Thimann. At the end of January, Reed set up a parallel Task Force, chaired by George Sprague for post-IBP internationally.

At the USEC/IBP meeting of February 15-16, both Fred Wagner and I argued that TIE was now interested in the initial establishment of a firm program base from which its interests and operations could be broadened at a later and more appropriate time and that the US/IBP biome studies represented that potential base in the post-IBP period. Following this discussion USEC/IBP decided that TIE should be invited to submit a proposal to US/IBP. A letter making this suggestion was sent by John Reed to Art Hasler on February 26.

The Task Forces met on May 1, 1973, just prior to the meeting of USEC/IBP, and formulated a set of recommendations which read:

"a. That based on information supplied by Dr. Blair, TIE be requested to assume coordination activities of the Biomes after June 1974 and to seek the necessary funding to continue the coordinating functions now handled through the Austin Office.

b. That NRC be asked to set up an exploratory panel, representing the array of disciplines required to consider the urban ecosystems problem and explore the possibilities of developing a specific proposal.

c. That the existing Biomes give greater attention to man in the Ecosystem.

d. That NRC be requested to establish a study panel to explore the need and feasibility of a biome approach to managed agriculture.

e. That Conservation of Ecosystems remain under the sponsorship of AIBS."

Hasler turned John Reed's letter over to Don Jameson, Chairman of the Board of Trustees of TIE, with the request that he serve as the interface between TIE and the IBP/EM Program Directors. The EM Program Directors met in February to discuss affiliation with TIE, and on March 16 Orie Loucks, as EM Coordinator, transmitted the results of their discussions to Don Jameson. Basically the desire was to find in TIE a substitute for the coordinating activities provided by the Austin office. Quoting from Loucks' letter:

"From the point of view of the directors of IBP ecosystem programs, there are several features that they see as essential in a successful sponsorship by TIE. The present programs represent years of effort in developing research teams that are rapidly advancing our understanding of diverse land and water systems, but whose goals will not have been met by the end of IBP. These features can be summarized as:

1. A commitment by TIE that it will support the existing ecosystem programs until they have had a reasonable opportunity to meet their research goals. This is primarily a concern that TIE not expect to move abruptly toward numerous new ecosystem programs, although this may be a desirable long-range goal. It is also understood that the ecosystem studies themselves have primary responsibility for developing support for their programs from federal or state agencies.

2. The directors also are desirous that TIE not embark immediately on large-scale competing programs (such as the TIE Laboratory), at least until the ecosystem studies program can be established and the new programs can be complementary rather than competitive."

The primary goals of an Office of Programs in Ecosystem Studies in TIE were summarized by Loucks as:

"1. Provision of a mechanism for inter-program coordination during the conclusion of the current integrated ecosystem programs (during post-IBP years), as well as stimulating development of new initiatives in inter-disciplinary study of ecosystems, building on the IBP experience;

2. Providing the means by which the principles of data exchange and inter-program use of data banks developed during the IBP can be managed in an equitable national context (within limits that protect the original investigator), for the benefit of the entire scientific community; and

3. Facilitating the exchange of concepts, models, and other advances from the various integrated research projects, and seeing that these advances are moved smoothly between institutions and the state and federal agencies, particularly as they relate to the evaluation of environmental impacts and solution of resource management problems."

Don Jameson replied to John Reed's letter on March 28, attempting to answer the questions raised in Reed's letter, but making no commitment concerning TIE-IBP affiliation. In fact, his response seemed negative:

"Sponsorship of US/IBP biomes. These programs are functioning and have developed institutional mechanisms consistent with their own needs; it is doubtful that there is any significant advantage for TIE to assume operation of existing individual biome programs as long as adequate scientific quality and problem focus can be maintained."

However, Jameson took a more positive position in a letter to Orie Loucks on April 12 saying that, "TIE does not necessarily seek sponsorship of IBP programs, but would welcome those programs (or components of programs) which do want to affiliate with or be operated by TIE."

Loucks summarized the state of negotiations, as reflected in the correspondence with TIE in a memorandum to me of April 24 as follows:

"• The Directors of ecosystem-related programs within the US.IBP are agreed that they would like to affiliate with TIE in a post-IBP relationship. This relationship would not preclude collaboration in other formats with on-going international programs, but it recognizes the mainly domestic mission on which these programs are focused.

"• The Institute of Ecology, with the concurrence of the Program Directors, is prepared to appoint a TIE Advisory Committee for Ecosystem Studies. The committee would advise and make recommendations to the TIE Directorate, and would be made up of the directors of various affiliated ecosystem programs, as well as others knowledgable of the need for ecosystem studies.

"• The Institute of Ecology is prepared to operate a program of Ecosystem Studies that might carry on some research not already underway and provide inter-program coordination and services needed by the TIE-affiliated ecosystem programs. The TIE-operated program of Ecosystem Studies would be headed by an Associate Director for Ecosystem Studies, responsible to the Director of TIE."

An affiliation agreement was drafted to be signed by IBP Program Directors and by Art Hasler on behalf of TIE. Two resolutions pertinent to this activity were passed by the TIE Trustees at their June 15, 1973 meeting. These were:

"*IBP-TIE Affiliation*: Taking note of the discussions on the previous day with IBP Program Directors, and the revised draft of an affiliation agreement between such programs and TIE, the Board
RESOLVED: to authorize the Director to execute the revised agreements and to forward them for consideration to the seven IBP programs party to these negotiations and to the Hubbard Brook Ecosystem Analysis program.

Noting that grant application for the EM Office's continuation would have to be submitted to NSF shortly, the Trustees

RESOLVED: to authorize the Director, in concert with the EM Office, to submit a proposal for the establishment of an Office of Ecosystem Studies within TIE, in accordance with the terms of the affiliation agreement."

The five biome directors, the Island Ecosystem director, and I as director of Origin and Structure of Ecosystems, signed the agreement and became members of the Committee on Ecosystem Studies (CES). A declination was received from the Hubbard Brook Project of Herb Bormann and Gene Likens, the only non-IBP project to receive an invitation to affiliate. However, the option to affiliate was kept open, and Likens who had been invited independently of his connection with Hubbard Brook accepted membership on the committee. Other non-IBP persons who were invited and accepted membership were Maximo Cerame-Vivas of the University of Puerto Rico, Arturo Gomez-Pompa of Universidad Nacional Autonoma de Mexico, R. Brooke Thomas of Cornell University, and Henry Regier of the University of Toronto.

The report on July 5 of the USNC/IBP Task Force on Post-IBP Programs and arrangements carried a recommendation that was supportive of what was already in process:

"The Task Force recommends:

That The Institute of Ecology (TIE) be requested to assume responsibility for coordination of the US/IBP Biome programs after June 1974, and that TIE seek necessary funding to continue the coordination functions now provided by the US/IBP Environmental Management Coordinating Office at the University of Texas, Austin."

Don Jameson asked me to convene the first meeting of CES. This meeting was held in Austin on September 10-11, 1973. One of our first tasks was to find a strong chairman, preferably one who had not been closely associated with the IBP. Subsequent to this meeting, we were able to convince Richard Parker of Washington State University that he should take the chairmanship.

A proposal for support of CES was made to NSF, with Orie Loucks as Project Director. This went from Art Hasler in January 1974 and was funded at the University of Wisconsin as of July 1. One of the terminal activities of US/IBP that found a home in TIE/CES and funding in the NSF grant to TIE for CES was the publication of the U.S. series of synthesis volumes presenting the results of the U.S. participation in the IBP other than those to be reported in the International Series sponsored by SCIBP and to be published by Cambridge University Press. Since the US/IBP had developed to a complexity of research on ecosystems not matched elsewhere in the IBP, the U.S. series promised to be a major undertaking and contribution.

The US/IBP Program Directors had appointed a Standing Committee on Publications, chaired by Bob Burgess of the Eastern Deciduous Forest Biome. This committee invited proposals from potential U.S. publishers of the U.S. series and eventually recommended that an agreement be signed with the relatively young firm of Dowden, Hutchinson, and Ross. A memorandum of agreement was drafted and eventually approved by TIE, legal counsel of NSF, and Dowden, Hutchinson, and Ross. I was asked to chair the Publications Committee and somewhat reluctantly accepted. In mid-1974 I was able to put together an excellent committee of Gabriel Lasker of Wayne State University, Bob Platt of Emory University, and Fred Smith of Harvard University.

6

Significance of the IBP

It is premature and probably presumptious at this time to try to judge the successes and failures of the IBP as an international effort in an area of science that is highly significant for the future existence of the world's human population at a tolerable level of environmental quality. In the United States in particular the IBP has been "big biology." Somewhere between 40 and 50 million dollars in Federal funds have been invested in the US/IBP, and it is impossible to estimate the financial investment of the 57 other participant countries. A large fraction of the environmentally oriented scientists of this country and of the world have invested a significant part of their research careers in the IBP. Powerful politicians and governmental agency people have solidly supported the IBP and by their support have made it financially and operationally feasible to carry out the IBP research.

It is appropriate now to address the question of whether these investments were worthwhile, even though final answers necessarily await forthcoming publication of the technical results of the IBP.

The significance of the IBP nationally and internationally must be treated separately, although the two interrelate to some considerable degree.

The IBP Internationally

As an experiment in international coordination and cooperation in science the IBP has had some conspicuous successes and some conspicuous failures. With respect to the latter, two major disappointments of the IBP as an international undertaking are outstanding.

Firstly, the IBP virtually failed to get beyond the "form-a-national committee" stage in the lesser developed countries (LDC's), where the IBP hopefully might have been the vehicle for the much needed advancement of the quality and quantity of environmental science. There are probably two main reasons for this failure. The most important is probably the lack of financial support. The second is probably that the relatively few scientists in most LDC's were just not conceptually ready for participation in cooperative, multidisciplinary or integrated research.

159

The IBP in Latin America is a case in point. Of the seven countries I visited in 1967 to promote participation in the IBP, there was stated enthusiasm in all for participation in the IBP, and a national committee was established in all 7, as well as in Ecuador, Panama, and Mexico. However, virtually no identifiable IBP research resulted in any of them other than the few binational projects in which the US/IBP took the initiative and involved local scientists in the cooperating country.

In retrospect, it seems that the scientists of the technologically advanced countries could have taken a more active role in goal-setting and in organizing IBP research to include the LDC's. Also there is no question in my mind but that a modest international fund to provide seed money for initiating IBP research in LDC's would have paid large dividends toward real involvement of the LDC scientists in this purportedly international effort.

The second major disappointment of the IBP internationally was the inability to generate a mechanism for the continued international coordination and cooperation in environmental science at the scientist-to-scientist, non-governmental level that has been one of the major accomplishments of the IBP.

As we discussed in detail in an earlier chapter, efforts to develop under ICSU and SCOPE a successor to SCIBP were fruitless in the face of relentless UNESCO opposition.

On the plus side of the ledger, the greatest accomplishment of the IBP internationally was the establishment of a previously unheard of level of communication, coordination, and cooperation among the world's environmental biologists as well as among its human biologists. This was accomplished by various means. Seven international task forces (called Sections) were formed and met regularly to coordinate and review activities under the seven major sub-themes of the IBP. Regular meetings of SCIBP, with broad international representation, were held to review the entire program. Five IBP congresses were held. Numerous international workshops were held to compare data and synthesize results. Some completely multinational research projects were mounted, although these were fewer than one might have hoped.

There seems little doubt that many of the lines of communications opened up among the scientists of diverse nations by the cooperation in the IBP will remain open. There seems little doubt, also, that the participants in the IBP effort now have a broadened perspective as regards the environmental issues facing us on this planet.

With regard to the advancement of ecological theory, it is premature to attempt to assess the contribution of the international activities of the IBP, as opposed to purely national efforts. My guess is that it will be considerable, especially where it has been possible to compare data on ecosystem processes on different continents and in different climatic regimes. However, a real assessment must await publication of the series of international synthesis volumes on IBP research now under way at Cambridge University Press.

Several current international activities have been influenced significantly by the ecosystem concepts advanced by the IBP. UNESCO's MAB was mentioned earlier. This was clearly patterned after the IBP. Whether or not MAB can succeed, where IBP largely failed, in mounting successful programs in which the LDC's will have major involvement, remains to be seen. Its intergovernmental nature presumably will make financial support more probable in the LDC's and in countries with closed political systems and rigid control of scientific institutions. How it will go in the United States remains to be seen.

In addition to UNESCO's MAB, several other significant international activities in the area of environmental science were initiated during the life of the IBP. These were both intergovernmental and non-governmental in scope. Persons deeply involved in the IBP were involved in the evolution of all of these, and the existence of the IBP had significant influence on all of them.

We have mentioned ICSU's SCOPE. There was very considerable overlap between SCIBP and SCOPE during the time that the latter was evolving its image and its mission. As it has evolved, SCOPE provides an international non-governmental body that, working through task forces, can assess the state of the art in various fields related to environmental management.

Another international activity under the umbrella of ICSU, and more specifically under one of its unions, IUBS (International Union of Biological Sciences) was the establishment of INTECOL (International Association for Ecology). Its main activity has been the holding of the first International Congress of Ecology in The Hague in September 1974. Persons from the IBP participated strongly in the planning, and the IBP organized five half-day symposia.

The potentially most significant international undertaking that emerged during the life of the IBP is the United Nations Environmental Program (UNEP) which resulted from the UN Conference on the Human Environment held in Stockholm in June 1972. Here, the impact of the IBP is clearly identifiable. Many IBP scientists participated in developing the voluminous work papers for the Conference. ICSU had a sizeable delegation of official observers at Stockholm with strong representation from SCOPE and IBP.

The imprint of the IBP on UNEP is clear in at least three aspects of that UN program:

(1) The global program that emerged from Stockholm is pervaded by the concept, clearly inherited from the IBP, of the ecosystem as the basis for environmental management.

(2) The global environmental monitoring program of UNEP is deeply rooted in an IBP activity, and subsequent SCOPE monitoring commission report which were discussed earlier. The report of the SCOPE commission played a major role in shaping the plan for global monitoring that emerged from Stockholm. My major concern today is that the concept of baseline monitoring of the health of representative ecosystems, a prominent part of the original plan, may have been lost.

(3) Thirdly, an IBP project for the conservation of plant gene pools was incorporated in the UNEP program thanks largely to the efforts of Sir Otto Frankel of Australia, one of the vice-presidents of SCIBP.

In my opinion UNEP is the one organization presently in existence that may have the capability of transforming the basic scientific knowledge of the structure and function and interrelatedness of the world's ecosystems into meaningful political action on a global scale. I believe that much of the future of this planet and of man on it depends on the ability to make this transfer from scientists to politicians and environmental developers and managers.

The Impact of the IBP in the United States

The impact of the IBP has varied greatly among the various participating countries. Probably no participant country has benefitted nationally to the extent that the United States has. It is my conviction that in the United States the IBP has been responsible for a quantum advance in ecological science as a basic science and for significant steps toward the development of the linkages necessary for the application of ecological knowledge to the solution of major environmental problems.

The IBP came along at a serendipitous time in the United States as well as in the whole world. In the United States in the 1960's there was mounting concern about pollution of air, water, and land and about general degradation of the environment. There was a large body of scientists who were willing to undertake ecological studies at a level of complexity (the biome) that was previously almost unthinkable. There were forward-thinking congressmen such as George Miller and Emilio Daddario, and others who were willing to carry the battle to obtain funding for this experiment in "big biology."

It is possible to cover virtually only in outline form the reasons why the IBP in the United States has triggered a major advance in the environmental science of ecology:

(1) The IBP has clearly demonstrated the feasibility of large integrated programs of research in environmental biology, involving up to hundreds of scientists of diverse disciplines, all working under a centrally managed project to obtain through their joint efforts an understanding of the structure and function of a highly complex ecological system. In the early days, when the IBP was struggling to get under way and to obtain funding, the statement was often heard that scientists would not submit to the kind of central control that would be necessary in an integrated research program (IRP). This came both from certain "lone-wolf" individuals among ecologists and from some non-ecologists who feared that the IBP would divert research funds from their own pet areas and were looking for any excuse to discredit it. As a matter of fact, the IBP has clearly demonstrated the fallacy of those arguments. Management practice has varied from that of central funding and subsequent subcontracting, as in the five biome studies, to separate grants to program components as in the Origin and Structure of Ecosystems Program. Both have proven feasible.

(2) The main achievement of the IBP in the United States has been the advancement of the concept of the ecosystem as a unit for ecological investigation. The thought emerging here is that the ecosystem provides a unit for management of the world's resources. In any particular ecosystem, with adequate knowledge of its structure and function, it should be possible to predict most reverberations through the system that result from any particular alteration of any component of the system.

(3) A third major accomplishment of the IBP has been the recognition of the fact that man is an integral and significant part of the ecosystem in which he occurs. Unfortunately, the early planning of the IBP involved a sharp distinction between man (Human Adaptability, HA Program) and the environmental (all other IBP programs). As the IBP progressed, the fallacy of this dichotomy became increasingly apparent. As the IBP draws to a close, there is a previously unheard of dialogue between human biologists and basic ecologists to the benefit of both disciplines.

(4) The IBP in the United States has been largely responsible for the genesis of a whole generation of ecosystem modellers, generally identifiable by their youth, their haircuts (long) and their attire (informal). Their efforts will have a lasting effect on the field of ecology.

(5) Also, a few of the projects in which U.S. scientists have participated successfully have demonstrated the feasibility, as well as the pitfalls to be avoided, of mounting operationally truly international or binational ecological research. These projects

include: (1) Man in the Andes, with Peruvian scientists, (2) South American Indians with Brazilian and Venezuelan cooperation, (3) Tundra Biome and Circumpolar Peoples, with various cooperating countries, (4) Origin and Structure of Ecosystems, with Argentine and Chilean scientists, (5) Upwelling Ecosystems with Latin American and European cooperation.

(6) Finally, there is, of course, the huge body of hard data on ecosystem processes that has appeared in the research literature and will continue to do so for years to come. In addition to this, the results of the US/IBP will appear in a series of some 20 synthesis volumes now in preparation. The impact of these results will speak for itself.

As the IBP closes down, it is my firm conviction that, here in the United States, the IBP has been far more successful than the most optimistic of its promoters had believed possible in the early days of struggle to get it accepted and under way. The IBP has catalyzed a major advance in the science of ecology. Furthermore, the emergent ecosystem ecology has tremendous potential as a basis for development and management of the biosphere's resources. It may not be too strong a statement to say that man's future depends on the integration into societal planning at all levels of government of the recognition of man as one component of an extremely complex megasystem concerning which we have now made modest steps toward understanding how this system is structured and how it functions.

My evaluation of the results of the IBP is clearly premature. Evidence for the success or failure of the big biology approach to ecosystem ecology will become convincing only as the 20 or more projected synthesis volumes reporting US/IBP results reach publication. Two other attempted evaluations of the US/IBP are equally premature. On of these (Committee to Evaluate the IBP, 1975) was a report of a committee appointed by NAS/NRC and chaired by Paul Kramer. This was a reasonably objective and fairly favorable assessment of the success of the US/IBP in achieving its objectives. However, an account of this report in *Science* (Boffey, 1976) was written seemingly to give the most negative possible view of IBP accomplishments. An earlier article in *Science* (Mitchell et al., 1976) was based on a study done for NSF by the Battelle Columbus Laboratories. This article contained such amazing errors as the statement that, "There still are no firm plans for the synthesis volumes." In concluded that, "integration so far has produced papers with only slight differences with respect to breadth of coverage and attention to dynamic relations than the average paper published in *Ecology*." This quite overlooks the fact that the main result of the integration will be seen in the synthesis volumes, which are just now being written and produced.

I will stick to my evaluation of the success of the IBP until better evidence to the contrary than is now available is produced. In this context we might quote a footnote from the NAS/NRC report which quotes A. R. Clapham, Chairman of the IBP National Committee for the United Kingdom as saying, "50 years hence people will point to important developments in biology and say, 'these are a direct result of the IBP.' "

References

Boffey, P. M. 1976. International Biological Program: Was it worth the cost and effort? Science 193:866-868.

Mitchel, R., R. A. Mayer, and J. Downhower. 1976. An evaluation of three biome programs. Science 192:859-865.

NAS/NRC Committee to Evaluate the IBP. 1975. An evaluation of the International Biological Program. 81 p. (Mimeo).

Stebbins, G. L. 1962a. Toward better international cooperation in the life sciences. Plant Sci. Bull. 8:105.

Stebbins, G. L. 1962b. International horizons in the life sciences. AIBS Bull. December: 13-19.

Worthington, E. B. 1975. The evolution of IBP. Cambridge Univ. Press.

166

Appendix A

Integrated Research Programs of the US/IBP

ANALYSIS OF ECOSYSTEMS

DIRECTORATE: Director, Frederick E. Smith, 1967-70; Committee of Biome Directors, 1970-74.

OBJECTIVES:
1. To understand how ecological systems operate with respect to both short-term and long-term processes.
2. To analyse interrelationships between land and water systems, so that broad regions may be considered as wholes.
3. To estimate existing and potential plant and animal production in the major climatic regions of this country, particularly in relation to human welfare, as our contribution to the worldwide goals of the IBP.
4. To add to the scientific basis of resource management, so that optimization for multiple use and for long-term use can be improved.
5. To establish a scientific base for programs to maintain or improve environmental quality.
6. To derive broad principles of ecosystem structure and function through an integration of the results of the Biome Studies.
7. To relate these principles to characteristics of ecosystems such as persistence, stability, maturity, and diversity.
8. To develop and refine a generalized adaptable simulation model suitable for use in planning studies for new development projects.
9. To elucidate productivity, nutrient cycling, energy flow, and other characteristics of ecosystems in a set of distinct environments.
10. To determine the driving forces, the processes causing transfers of matter and energy among

components, the non-concentration characteristics, and the controlling variables in each Biome.

11. To determine the ecosystem response to the natural and man-induced stresses appropriate to each Biome; for example: large herbivores in grassland, extreme and rare weather patterns in desert, periodic fluctuations of rodent populations in tundra, commercial use of timber in the coniferous forests, and urbanization in the deciduous forest.

12. To understand the land-water interactions characteristic of each Biome: prairie ponds and reservoirs in the grasslands, the abundance of shallow waters in wet tundra, springs and temporary waters in deserts, river systems with anadromous fish populations in coniferous forest, and pollution and eutrophication in the deciduous forest.

13. To synthesize the results of these and previous studies into predictive models of temporal and spatial variation, effects of pollutants and of exploitation, stability, and other ecosystem characteristics necessary for resource management in each Biome.

Grassland Biome

DIRECTORATE: Director, George M. Van Dyne, 1967-74; Director, James H. Gibson, 1974-76

RESEARCH SITES:

Principal Site: Pawnee (Central Plains Experimental Range of USDA Agricultural Research Service) about 30 miles east of Fort Collins, Colorado.

Other Sites: ALE (Hanford), Washington
Bison, Montana
Bridger, Montana
Cottonwood, South Dakota
Jornada, New Mexico
Osage, Oklahoma
Pantex, Texas
San Joaquin, California

DURATION OF PROGRAM: 1967-76

OBJECTIVES:

To gain understanding of the structure and functioning of grassland ecosystems.

RESULTS:

A. To October 1976, 383 (including abstracts) or 271 (without abstracts) papers in open literature, 303 technical reports, 11 Range Science Series, 61 M.S. theses, and 46 Ph.D. dissertations.

B. Series of grassland type volumes in book form in preparation:
   1. Structure, Function and Utilization of the Annual Grassland Ecosystem. Don A. Duncan, Robert G. Woodmansee.
   2. The True Prairie Ecosystem. Paul G. Risser.
   3. Structure, Function, and Utilization of Mixed Prairie Ecosystems. James K. (Tex) Lewis.
   4. Structure, Function, and Utilization of North American Desert Grassland Ecosystems. Rex D. Pieper.
   5. The Pacific Northwest Shrub-Steppe Ecosystem. William H. Rickard.
   6. The Shortgrass Prairie Ecosystem. Norman R. French, Freeman M. Smith, George M. Van Dyne.
   7. Mountain Grasslands of the North American Rocky Mountains. Tad Weaver.

C. North American Grassland Synthesis Volume

PARTICIPANTS
Primary Productivity—Plants
Bartos, Dale E., Colorado State University
Begg, Eugene L., University of California, Davis
Beidleman, Richard G., Colorado College
Bement, Robert E., ARS, USDA, Fort Collins, Colorado
Bokhari, Unab G., Colorado State University
Burcham, L. T., Sacramento, California
Caldwell, Martyn M., Utah State University
Cline, Jerry F., Battelle Memorial Institute, Washington
Collins, Don D., Montana State University
Davidson, Darwin, University of Wyoming
Detling, James K., Colorado State University
Dix, Ralph L., Colorado State University
Dodd, Jerrold L., Colorado State University
Duncan, Don A., U.S. Forest Service, Fresno, California
Evans, Raymond A., University of California, Berkeley
Fisser, Herbert G., University of Wyoming
Gilbert, Richard O., Battelle Memorial Institute, Washington
Heady, Harold F., University of California, Berkeley
Herbel, Carlton H., ARS, USDA, Las Cruces, New Mexico
Hulett, Gary K., Fort Hays Kansas State College
Hutcheson, Harvie L., Jr., South Dakota State University
Hyder, Donald N., ARS, USDA, Fort Collins, Colorado
Jain, Subodh K., University of California, Davis
Jones, Milton B., University of California, Hopland
Kay, Burgess L., University of California, Davis
Kennedy, Robert K., University of Oklahoma
Klein, William M., Colorado State University
Knight, Dennis H., University of Wyoming
Knievel, Daniel P., University of Wyoming
Lauenroth, William K., Colorado State University
Lewis, James K., South Dakota State University
Marshall, John K., Colorado State University
McColl, John G., South Dakota State University
Menke, John W., University of California, Berkeley
Miller, Lee D., Colorado State University
Moir, William H., Colorado State University
Morris, Melvin S., University of Montana
Pettit, Russell D., Texas Tech University
Pieper, Rex D., New Mexico State University
Raguse, Charles A., University of California, Davis
Rickard, William H., Jr., Battelle Memorial Institute, Washington
Risser, Paul G., University of Oklahoma

Shushan, Sam, University of Colorado
Singh, J. S., Colorado State University
Street, James E., University of California, Davis
Terwilliger, Charles, Jr., Colorado State University
Tomanek, George W., Fort Hays Kansas State College
Trlica, M. Joseph, Colorado State University
Uresk, Daniel W., Battelle Memorial Institute, Washington
Ward, Richard T., Colorado State University
Weaver, Theodore W., Montana State University
Whitman, Warren C., North Dakota State University
Williams, George J. III, Washington State University
Williams, William A., University of California, Davis
Young, James A., University of Nevada

Secondary Productivity—Animals
Andrews, Robin M., Colorado State University
Baldwin, Paul H., Colorado State University
Bell, Ross T., University of Vermont
Bement, Robert E., ARS, USDA, Fort Collins, Colorado
Birney, Elmer C., University of Minnesota
Blocker, H. Derrick, Kansas State University
Burdick, Donald J., California State University, Fresno
Clawson, W. James, University of California, Davis
Coleman, David C., Colorado State University
Cringan, Alexander T., Colorado State University
Crossley, D. A., Jr., University of Georgia
Dean, Ronald E., University of Wyoming
Dickinson, Charles E., Colorado State University
Dyer, Melvin I., Colorado State University
Ellis, James E., Colorado State University
Ells, James O., Denver Wildlife Research Center
Fitzner, R. E., Battelle Memorial Institute, Washington
French, Norman R., Colorado State University
Gerdemann, James W., University of Illinois
Gies, R. A., Battelle Memorial Institute, Washington
Gilbert, Richard O., Battelle Memorial Institute, Washington
Golley, Frank B., University of Georgia
Graul, Walter D., University of Minnesota
Gross, Jack E., Fish and Wildlife Service, Fort Collins, Colorado
Grow, Russell R., University of Wyoming
Hansen, Richard M., Colorado State University
Harris, Lawrence D., University of Florida
Hedlund, John D., Battelle Memorial Institute, Washington
Hoffmann, Robert S., University of Kansas
Huddleston, Ellis W., New Mexico State University

Hyder, Donald N., ARS, USDA, Fort Collins, Colorado
Jones, J. Knox, Jr., University of Kansas
Johnston, Donald E., Ohio State University
Koong, L. J., University of California, Davis
Koranda, John J., University of California, Livermore
Kumar, Rabinder, University of Wyoming
Lavigne, Robert J., University of Wyoming
Leetham, John W., Colorado State University
Lloyd, John E., University of Wyoming
McDaniel, Burruss, Jr., South Dakota State University
McEwen, Lowell C., Denver Wildlife Research Center
Mitchell, John E., Colorado State University
Nagy, Julius G., Colorado State University
O'Brien, Charles W., Texas Tech University
O'Farrell, Thomas P., Battelle Memorial Institute, Washington
Olendorff, Richard R., American Museum of Natural History, Washington, D.C.
Packard, Robert L., Texas Tech University
Paris, Oscar H., University of Wyoming
Paur, Leonard F., Colorado State University
Pfadt, Robert E., University of Wyoming
Pigg, Joanne, Fresno State University, California
Raitt, Ralph J., New Mexico State University
Rice, Richard W., University of Wyoming
Rogers, Lee E., University of Wyoming
Ryder, Ronald A., Colorado State University
Savic, I. R., University of Georgia
Schitoskey, Frank, Jr., South Dakota State University
Scott, James A., Colorado State University
Sims, Phillip L., Colorado State University
Swift, David M., Colorado State University
Thatcher, Theodore O., Colorado State University
Thomas, Bert O., University of Northern Colorado
Torell, Donald T., University of California, Hopland
Ueckert, Darrel N., Texas Tech University
Van Horn, Donald H., University of California
Wallace, Joe D., New Mexico State University
Ward, Charles R., Texas Tech University
Watts, John G., New Mexico State University
Wiens, John A., Oregon State University
Wunder, Bruce A., Colorado State University

Decomposition
Angel, Kathleen, La Roche College, Pittsburgh, Pennsylvania
Baugh, Clarence, Texas Tech University

Boulette, Ernest P. III, Texas Tech University
Buschbom, R. L., Battelle Memorial Institute, Washington
Christensen, Martha, University of Wyoming
Clark, Francis E., ARS, USDA, Fort Collins, Colorado
Coleman, David C., Colorado State University
Doxtader, Kenneth G., Colorado State University
Garland, T. R., Battelle Memorial Institute, Washington
Gorden, Robert W., Texas Tech University
Harris, John O., Kansas State University
Harris, Lawrence D., Colorado State University
Klein, Donald A., Colorado State University
Kucera, Clair L., University of Missouri
Mayeux, Jerry V., Colorado State University
McClellan, J. Forbes, Colorado State University
Pengra, Robert M., South Dakota State University
Porter, Ralph C., Texas Tech University
Schmidt, R. L., Battelle Memorial Institute, Washington
Smolik, James D., South Dakota State University
Staffeldt, Eugene E., New Mexico State University
Stanton, Nancy L., University of Wyoming
Thayer, Donald W., Texas Tech University
Wicklow, Donald T., University of Pittsburgh
Wildung, Raymond E., Battelle Memorial Institute, Washington

Hydrology
Becker, Clarence F., University of Wyoming
Burman, Robert D., University of Wyoming
Burzlaff, Donald F., University of Nebraska
Franklin, William T., Colorado State University
Hanson, Clayton L., ARS, USDA, South Dakota
Hasfurther, Victor R., University of Wyoming
Hein, Dale, Colorado State University
Herrmann, Scott J., Southern Colorado State College
Kanemasu, Ed, Kansas State University
Knowlton, Dennis J., University of Wyoming
Koranda, John J., University of California, Livermore
LaVelle, James W., Southern Colorado State College
Marlatt, William E., Colorado State University
Nunn, John R., University of Wyoming
Price, Keith, Battelle Memorial Institute, Washington
Rauzi, Frank, ARS, USDA, Wyoming
Seilheimer, Jack A., Southern Colorado State College
Smith, Freeman M., Colorado State University
Striffler, W. David, Colorado State University

Trappe, James, Forest Service, Corvallis, Oregon
Weeks, Richard W., University of Wyoming

Meteorology
Armijo, Joseph D., University of Wyoming
Becker, Clarence F., University of Wyoming
Burman, Robert D., University of Wyoming
Knowlton, Dennis J., University of Wyoming
Koranda, John J., University of California, Livermore
Marlatt, William E., Colorado State University
Nunn, John R., University of Wyoming
Parton, William J., Jr., Colorado State University
Pochop, Larry D., University of Wyoming
Rasmussen, James L., Colorado State University
Smith, Freeman M., Colorado State University
Twitchell, George, University of Wyoming
Vandenberg, Robert P., Colorado State University
Weeks, Richard W., University of Wyoming
Whicker, F. Ward, Colorado State University

Nitrogen Cycling
Bowman, Rudolph A., Colorado State University
Burzlaff, Donald F., University of Nebraska
Clark, Francis E., ARS, USDA, Fort Collins, Colorado
Franklin, William T., Colorado State University
Kline, Jerry L., Argonne National Laboratory, Illinois
Porter, Lynn K., ARS, USDA, Fort Collins, Colorado
Reuss, John O., Colorado State University
Trappe, James, Forest Service, Corvallis, Oregon
Whicker, F. Ward, Colorado State University
Woodmansee, Robert G., Colorado State University

Phosphorus Cycling
Bowman, Rudolph A., Colorado State University
Burzlaff, Donald F., University of Nebraska
Cole, C. Vernon, ARS, USDA, Fort Collins, Colorado
Franklin, William T., Colorado State University
Porter, Lynn K., ARS, USDA, Fort Collins, Colorado
Trappe, James, Forest Service, Corvallis, Oregon
Woodmansee, Robert G., Colorado State University

Modeling
Anway, Jerry C., Colorado State University
Bledsoe, L. J., Colorado State University
Clymer, A. Ben, Ohio PEA

Cole, George W., Colorado State University
Connor, David J., La Trobe University, Australia
Detling, James K., Colorado State University
Francis, Robert C., Colorado State University
Gustafson, Jon D., Colorado State University
Hunt, William H., Colorado State University
Innis, George S., Colorado State University
Olson, Richard J., Battelle Memorial Institute, Washington
Parton, William J., Jr., Colorado State University
Patten, Bernard C., University of Georgia
Paur, Leonard F., Colorado State University
Rodell, Charles F., Colorado State University
Sauer, Ronald H., Colorado State University
Singh, J. S., Colorado State University
Smith, Freeman M., Colorado State University
Steinhorst, R. Kirk, Colorado State University
Swartzman, Gordon L., Colorado State University
Swift, David M., Colorado State University
Van Dyne, George M., Colorado State University
Wiens, John A., Oregon State University
Woodmansee, Robert G., Colorado State University

Central Administration, Services, and Advisory
Baker, C. Van, Colorado State University
Bigelow, David S., Colorado State University
Brooks, Alan, Colorado State University
Campion, Marilyn K., Colorado State University
Carlson, Barbara J., Colorado State University
Dyer, Melvin I., Colorado State University
French, Norman R., Colorado State University
Gibson, James H., Colorado State University
Gustafson, Jon D., Colorado State University
Haug, Peter T., Colorado State University
Innis, George S., Colorado State University
Jameson, Donald A., Colorado State University
Johnson, Corrine L., Colorado State University
Kanode, Mary L., Colorado State University
Keith, Vicki E., Colorado State University
Nell, Larry G., Colorado State University
Peltz, Jerry, Colorado State University
Rice, Wilburn A., Colorado State University
Robinson, Robert D., Colorado State University
Shipley, Richard L., Colorado State University
Smith, Freeman M., Colorado State University
Steinhorst, R. Kirk, Colorado State University

Stevens, Kim L., Colorado State University
Streeter, Charles L., Colorado State University
Swift, David M., Colorado State University
Van Dyne, George M., Colorado State University
Wright, R. Gerald, Jr., Colorado State University

External Advisors
Botkin, Daniel B., Yale University, Connecticut
Clymer, A. Ben, Ohio EPA
Daubenmire, Rexford, Washington State University
Delwiche, Connie C., University of California, Davis
Ellis, Robert H., Rensselaer Polytechnic Institute, New York
Kadlec, John A., Utah State University
Kline, Jerry L., Argonne National Laboratory, Illinois
Likens, Gene E., Cornell University, New York
Watt, Kenneth E. F., University of California, Davis
Wiegert, Richard G., University of Georgia

Site Administration
Collins, Don D., Montana State University
Dodd, Jerrold L., Colorado State University
Duncan, Don A., U.S. Forest Service, California
Huddleston, Ellis W., New Mexico State University
Jameson, Donald A., Colorado State University
Lewis, James K., South Dakota State University
Morris, Melvin S., University of Montana
O'Farrell, Thomas P., Battelle Memorial Institute, Washington
Pieper, Rex D., New Mexico State University
Rickard, William H., Jr., Battelle Memorial Institute, Washington
Risser, Paul G., University of Oklahoma
Shoop, Marvin C., Colorado State University
Souther, Raymond C., Colorado State University
Tomanek, George W., Fort Hays Kansas State College
Weaver, Theodore W., Montana State University
Whitman, Warren C., North Dakota State University

Visiting Scientists
  Primary productivity—plants
Ares, Jorge O., University of Buenos Aires, Argentina
Brittain, Edward G., Australian National University, Australia
Connor, David J., La Trobe University, Australia
Denisiuk, Zygmunt, Polish Academy of Sciences, Poland
Haydock, K. Paul, CSIRO, Australia
Thorsteinsson, Ingvi, University of Reykjavik, Iceland

Van Wyk, Jack J. P., Potchefstroom University, South Africa
Wielgolaski, Franz E., University of Oslo, Norway
    Secondary productivity—animals
Addison, Ken B., Brian Pastures, Australia
Barrett, Gary W., Miami University, Ohio
Breymeyer, Alicja, Polish Academy of Sciences, Poland
Egunjobi, Kola, University of Ibadan, Nigeria
Framstad, Erik, University of Oslo, Norway
Grodzinski, Wladyslaw, Jagiellonian University, Poland
Johanningsmeier, Arthur G., Cushing Academy, Massachusetts
Lebreton, Jean-Dominique, University of Lyon, France
    Decomposition
Clarholm, Marianne, Royal Agricultural College of Sweden, Sweden
    Modeling
Addison, Ken B., Brian Pastures, Australia
Brittain, Edward G., Australian National University, Australia
Connor, David J., La Trobe University, Australia
Framstad, Erik, University of Oslo, Norway
Haydock, K. Paul, CSIRO, Australia
Thorsteinsson, Ingvi, University of Reykjavik, Iceland
White, E. Graeme, Lincoln College, New Zealand
Wielgolaski, Franz E., University of Oslo, Norway

Graduate Students (Funded all or in part)
Abramsky, Zvika, Colorado State University
Alldredge, A. William, Colorado State University
Almeyda, Guillermo F., Colorado State University
Armijo, Joseph D., University of Wyoming
Armstrong, David M., Fort Hays Kansas State College
Bartos, Dale L., Colorado State University
Bauerle, Bruce A., University of Northern Colorado
Bertolin, G., Colorado State University
Betz, Sharon R., Colorado State University
Bhatnagar, Kailash N., University of Wyoming
Bledsoe, L. J., Colorado State University
Boyd, Roger L., Colorado State University
Brown, Larry F., Colorado State University
Brun, Lynn J., Kansas State University
Brunner, J. David, University of Montana
Butterfield, Joe D., Colorado State University
Cavender, Barbara R., Colorado State University
Chu, J., University of Wyoming
Chuculate, Charles A., Colorado State University
Cobb, D. A., Colorado State University

Cockerham, William, California State University, Fresno
Conley, Walter H., Texas Tech University
Connaughton, Martin, New Mexico State University
Copley, Paul W., Colorado State University
Coughenour, Michael B., Colorado State University
Creighton, Phillip D., Colorado State University
Crooks, Patti, University of Missouri
Cwik, Michael J., Colorado State University
Czaplewski, Raymond L., Colorado State University
Dearden, Boyd L., Colorado State University
Deblinger, Robert D., Colorado State University
Dittberner, Phillip L., Colorado State University
Donoho, Harvey S., Colorado State University
Dyck, George W., Colorado State University
Dye, A. J., Colorado State University
Eklund, Leslie R., Colorado State University
Ellstrom, Mark A., New Mexico State University
Fagan, Richard E., Texas Tech University
Fitzenrider, Robert, New Mexico State University
Flake, Lester D., Washington State University
Flinders, Jerran T., Colorado State University
Fournier, Walter J., Texas Tech University
Fraley, Leslie, Jr., Colorado State University
Fullerton, Sue A., Colorado State University
Galbraith, Alan F., Colorado State University
Garratt, Michael, Colorado State University
Giezentanner, J. Brent, Colorado State University
Gilbert, Bradley J., Colorado State University
Grant, William E., Colorado State University
Green, James L., Colorado State University
Grow, Russell R., University of Wyoming
Haglund, B., Montana State University
Hamm, Paula E., South Dakota State University
Haug, Peter T., Colorado State University
Helsel, Edith D., University of Pittsburgh
Hobbs, N. Thompson, Colorado State University
Hoover, John P., Colorado State University
Hughes, Joseph H., Colorado State University
Hutcheson, James L., University of Wyoming
Inyamah, Grace C., Colorado State University
Johnson, Forrest L., University of Oklahoma
Joyce, Linda A., Colorado State University
Keith, Edward O., Colorado State University
Kelly, James M., Colorado State University

Kerr, Stephen M., Colorado State University
Kisiel, Donald S., Colorado State University
Kirchner, Thomas B., Colorado State University
Lampe, Richard P., University of Kansas
Lauenroth, William K., Colorado State University
Lengkeek, Venance H., South Dakota State University
Lotze, Jorge H., Colorado State University
Marion, Wayne R., Colorado State University
Markley, J. L., University of Denver
Marti, Carl D., Jr., Colorado State University
Martin, Robert E., Texas Tech University
Mason, G. E., Kansas State University
Matocha, Kenneth G., Texas Tech University
May, Silas W., University of Oklahoma
Menapace, Diana M., Colorado State University
Menke, John W., Colorado State University
Miller, Janet L. S., Colorado State University
Miskimins, Richard D., Colorado State University
Mitchell, Glen C., University of Wyoming
Mitchell, John E., Colorado State University
Nunn, John R., University of Wyoming
Nyhan, John W., Colorado State University
Oftedahl, Larry O., Colorado State University
Olendorff, Richard R., Colorado State University
Parton, William J., University of Oklahoma
Pearson, Robert L., Colorado State University
Peden, Donald G., Colorado State University
Pefaur, Jaime E., University of Kansas
Pendleton, Dennis F., Colorado State University
Pimm, Stuart L., New Mexico State University
Porter, David K., Colorado State University
Pries, Robert, University of Missouri
Proctor, Charles W., Jr., University of Georgia
Redetzke, Keith A., Colorado State University
Reed, Rodman C., Kansas State University
Reid, Loren D., Colorado State University
Rodgers, David W., South Dakota State University
Rogers, Lee E., University of Wyoming
Rotenberry, John T., Oregon State University
Scarborough, Arla M., University of Wyoming
Schwartz, Charles C., Colorado State University
Seaman, Sally L., Colorado State University
Schmer, David H., University of Wyoming
Shaver, J. C., University of Wyoming
Shaw, Richard A., Jr., Texas Tech University

180   W. F. Blair

Sheedy, J., University of Oklahoma
Simpson, Timothy, Colorado State University
Smith, Freeman M., Colorado State University
Smolik, James D., South Dakota State University
Sparrow, Elena Bautista, Colorado State University
Stepanich, R. M., Kansas State University
Stout, William R., Colorado State University
Street, Roland, Texas Tech University
Strong, Mark A., Colorado State University
Stroud, Donald W., University of Wyoming
Sundberg, Eric S., Colorado State University
Tucker, C. J. III, Colorado State University
Uresk, Daniel W., Colorado State University
Van Haveren, Bruce P., Colorado State University
Vavra, Martin, University of Wyoming
Walters, Carl J., Colorado State University
Welch, W. R., Colorado State University
Wesley, David E., Colorado State University
Whitmill, Michael, Texas Tech University
Wiley, Robert W., Texas Tech University
Williams, Kenneth, Colorado State University
Wolff, Delbert N., Colorado State University
Woodmansee, Robert G., Colorado State University
Wright, R. Gerald, Jr., Colorado State University
Yount, Virginia A., Colorado State University
Zimmerman, U. Douglas, University of Missouri

Other
Currie, Pat O., Rocky Mountain Forest and Range Experiment Station, Fort Collins, Colorado
Forcella, F., Montana State University
Free, J. C., Colorado State University
Gibbens, Robert P., Range and Livestock Research Station, Cottonwood, South Dakota
Hervey, D. F., Colorado State University
Klett, Hollis, Texas Tech University
Laycock, William A., Rocky Mountain Forest and Range Experiment Station, Fort Collins, Colorado
Matthews, Samuel, Colorado State University
Morris, Meredith J., Rocky Mountain Forest and Range Experiment Station, Fort Collins, Colorado
Mueggler, Walter F., Intermountain Forest and Range Experiment Station, Bozeman, Montana
Murphy, Alfred H., University of California, Hopland

Paulsen, Harold A., Rocky Mountain Forest and Range Experiment
    Station, Fort Collins, Colorado
Price, Keith, Battelle Memorial Institute, Washington
Rosmond, Thomas, Environmental Prediction Research Facility, Naval
    Postgraduate School, Monterey, California
Smith, James A., Colorado State University
Sparks, Kelly L., Colorado State University
Travis, Michael, Colorado State University
Warren, C. Gerald, Colorado State University

Eastern Deciduous Forest Biome

DIRECTORATE: Director, Stanley I. Auerbach; Deputy Director, Robert L. Burgess

Executive Committee:
    Auerbach, Stanley I., Chairman, Oak Ridge National Laboratory
    Adams, Michael S., University of Wisconsin
    Burgess, Robert L., Oak Ridge National Laboratory
    Clesceri, Nicholas L., Rensselaer Polytechnic Institute
    Goff, F. Glenn, Oak Ridge National Laboratory
    Monk, Carl D., University of Georgia
    O'Neill, Robert V., Oak Ridge National Laboratory
    Reichle, David E., Oak Ridge National Laboratory
    Strain, Boyd R., Duke University
Advisory Committee:
    Bay, Roger, U.S. Forest Service
    Cooper, William, Michigan State University
    Kadlec, John A., University of Michigan
    Madgwick, Herbert A., Virginia Polytechnic Institute
    Neuhold, John M., Utah State University
    Wenger, Karl F., U.S. Forest Service

RESEARCH SITES:
    Principal Site:
      Oak Ridge National Laboratory, Tennessee
    Other Sites:
      Coweeta (Hydrological Laboratory, U.S. Forest Service), Franklin, North Carolina
      Lake George, Troy, New York
      Lake Wingra, Madison, Wisconsin
      Triangle, Durham, North Carolina

DURATION OF PROGRAM: 1969-75

OBJECTIVES: To gain understanding of the structure and functioning of deciduous forest ecosystems.

RESULTS: 262 papers in open literature (to June 1976), one synthesis volume ("Ecosystem Analysis in the Eastern Deciduous Forest") in US/IBP series (in preparation).

Eastern Deciduous Forest Biome

PARTICIPANTS

NAME	INSTITUTION
Adams, Michael S.	University of Wisconsin
Allen, Timothy	University of Wisconsin
Anderson, Roger C.	University of Wisconsin
Armstrong, D. E.	University of Wisconsin
Art, Henry W.	Williams College
Auerbach, Stanley I.	Oak Ridge National Laboratory
Aulenbach, Donald B.	Rensselaer Polytechnic Institute
Ausmus, Beverly S.	Oak Ridge National Laboratory
Bay, Roger	U.S. Forest Service
Baxter, F. Paul	Oak Ridge National Laboratory
Best, G. Ronnie	University of Georgia
Best, Sam	University of Georgia
Bloomfield, Jay	Rensselaer Polytechnic Institute
Bogdan, Kenneth	State University of New York, Albany
Booth, Ray S.	Oak Ridge National Laboratory
Boyle, James Reid	University of Wisconsin
Boylen, Charles	Rensselaer Polytechnic Institute
Buol, Stanley W.	North Carolina State University
Burgess, Robert L.	Oak Ridge National Laboratory
Burnette, Carroll	U.S. Forest Service
Buzzard, Gale	Duke University
Chapman, Roger	Duke University
Chisolm, Sallie	State University of New York, Albany
Churchill, Warren	Wisconsin Department of Natural Resources
Clebsch, Edward E. C.	University of Tennessee
Clesceri, Lenore S.	Rensselaer Polytechnic Institute
Clesceri, Nicholas L.	Rensselaer Polytechnic Institute
Colon, Emilio	Rensselaer Polytechnic Institute
Comiskey, Charles	Oak Ridge National Laboratory
Cooper, William	Michigan State University
Cornaby, B. W.	University of Georgia
Cottam, Grant	University of Wisconsin

PROGRAM AFFILIATION	ACTIVITY
Lake Wingra Site	Site Coordinator
Biome and Regional Analysis	
Biome and Regional Analysis	
Lake Wingra Site	Aquatic Mineral Cycling
Biome and Regional Analysis	
Management	Director
Lake George Site	Aquatic Mineral Cycling
Oak Ridge Site	Terrestrial Decomposition
Management	Advisory Committee
Biome and Regional Analysis	
Coweeta Site	Terrestrial Decomposition
Coweeta Site	Terrestrial Mineral Cycling
Lake George Site	Modeling
Lake George Site	Aquatic Primary Production
Modeling	
Lake Wingra Site	Terrestrial Mineral Cycling
Lake George Site	Aquatic Decomposition
Triangle Site	Terrestrial Mineral Cycling
Management	Deputy Director
Triangle Site	Terrestrial Mineral Cycling
Triangle Site	Modeling
Triangle Site	Terrestrial Primary Production
Lake George Site	Aquatic Decomposition
Lake Wingra Site	Aquatic Secondary Production
Biome and Regional Analysis	
Lake George Site	Aquatic Decomposition
	Process Coordinator
Lake George Site	Site Coordinator
Lake George Site	Hydrology
Oak Ridge Site	Land/Water Interaction
Management	Advisory Committee
Coweeta Site	Terrestrial Decomposition
Biome and Regional Analysis	

NAME	INSTITUTION
Cox, Tracy L.	Oak Ridge National Laboratory
Crossley, D. A., Jr.	University of Georgia
Crow, Thomas R.	U.S. Forest Service
Dahlman, Roger C.	Oak Ridge National Laboratory
Dase, Michel	Rensselaer Polytechnic Institute
Day, F. P., Jr.	University of Georgia
DeAngelis, Donald L.	Oak Ridge National Laboratory
Deselm, Hal R.	University of Tennessee
Dettman, Edward H.	University of Wisconsin
Dinger, Blaine E.	Oak Ridge National Laboratory
Douglass, James	U.S. Forest Service
Dreher, Frederick	U.S. Geological Survey
Dueser, Raymond	Oak Ridge National Laboratory
Dunn, E. Lloyd	University of Georgia
Edwards, Nelson T.	Oak Ridge National Laboratory
Ek, Alan R.	University of Wisconsin
Elwood, Jerry W.	Oak Ridge National Laboratory
Endelman, Fred	University of Wisconsin
Ensign, Jerald C.	University of Wisconsin
Embry, Richard	University of Georgia
Ferguson, Nancy	Oak Ridge National Laboratory
Ferris, James J.	Rensselaer Polytechnic Institute
Ferris, John M.	Purdue University
Fisher, John	Rensselaer Polytechnic Institute
Franz, Eldon H.	University of Georgia
Gasith, Avital	University of Wisconsin
Gist, Clayton	University of Georgia
Goff, F. Glenn	Oak Ridge National Laboratory
Goldstein, Robert A.	Oak Ridge National Laboratory
Gonthier, Joseph B.	U.S. Geological Survey
Green, Theodore, III	University of Wisconsin
Gregory, Francis	Rensselaer Polytechnic Institute
Gresham, Charles	Duke University
Grizzard, Thomas	Oak Ridge National Laboratory
Harris, Robin F.	University of Wisconsin
Harris, W. Frank	Oak Ridge National Laboratory
Hasler, Arthur D.	University of Wisconsin
Hellmers, Henry	Duke University

PROGRAM AFFILIATION	ACTIVITY
Oak Ridge Site	Terrestrial Mineral Cycling
Coweeta Site	Site Coordinator
	Process Coordinator
Biome and Regional Analysis	
Oak Ridge Site	Terrestrial Mineral Cycling
Lake George Site	Aquatic Decomposition
Coweeta Site	Terrestrial Primary Production
Management	Modeling
Biome and Regional Analysis	
Lake Wingra Site	Modeling
Oak Ridge Site	Terrestrial Primary Production
Coweeta Site	Hydrology
Lake Wingra Site	Hydrology
Oak Ridge Site	Terrestrial Secondary Production
Coweeta Site	Terrestrial Primary Production
Oak Ridge Site	Terrestrial Decomposition
Biome and Regional Analysis	
Oak Ridge Site	Aquatic Secondary Production
Lake Wingra Site	Aquatic Mineral Cycling
Lake Wingra Site	Aquatic Decomposition
Coweeta Site	Meteorology
Management	Information Center Coordinator
Lake George Site	Aquatic Decomposition
Lake Wingra Site	Terrestrial Decomposition
Lake George Site	Information Center
Coweeta Site	Terrestrial Primary Production
Lake Wingra Site	Aquatic Mineral Cycling
Coweeta Site	Modeling
Biome and Regional Analysis	Coordinator
Management	Modeling
Lake Wingra Site	Hydrology
Lake Wingra Site	Hydrology
Lake George Site	Aquatic Mineral Cycling
Triangle Site	Terrestrial Primary Production
Oak Ridge Site	Land/Water Interaction
Lake Wingra Site	Land/Water Interaction
Oak Ridge Site	Site Coordinator
	Terrestrial Primary Production Process Coordinator
Lake Wingra Site	Aquatic Secondary Production
Triangle Site	Terrestrial Primary Production

NAME	INSTITUTION
Henderson, Gray S.	Oak Ridge National Laboratory
Hesketh, John D.	Agricultural Research Service
Higginbotham, Kenneth	Duke University
Hilsenhoff, William L.	University of Wisconsin
Hole, Francis D.	University of Wisconsin
Holt, Lee	U.S. Geological Survey
Hoopes, John A.	University of Wisconsin
Howard, H. H.	Skidmore College
Howell, Evelyn	University of Wisconsin
Huff, Dale D.	Oak Ridge National Laboratory
Hunter, Preston	University of Georgia
Hutchison, Boyd	Atmospheric Turbulence and Diffusion Laboratory, NOAA
Jenkins, Van	North Carolina State University
Johnson, Mary Lee	Oak Ridge National Laboratory
Johnson, W. Carter	Oak Ridge National Laboratory
Jorgensen, Jacques R.	U.S. Forest Service
Kadlec, John A.	University of Michigan
Katz, Sam	Rensselaer Polytechnic Institute
Keeney, Dennis R.	University of Wisconsin
Kercher, James	Oak Ridge National Laboratory
Kiggins, Russ	Rensselaer Polytechnic Institute
Kinerson, Russell	Duke University
Kitchell, James R.	University of Wisconsin
Kleiber, William	University of Wisconsin
Kobayashi, Shigeru	Rensselaer Polytechnic Institute
Kobriger, Nick	University of Wisconsin, Milwaukee
Kohberger, Bob	Rensselaer Polytechnic Institute
Koonce, Joseph	University of Wisconsin
Knoerr, Kenneth R.	Duke University
LaRow, Edward J.	Siena College
Lettau, Bernard	State University of New York, Albany
Lieth, Helmut	University of North Carolina
Loucks, Orie L.	University of Wisconsin
Luxmoore, Robert	Oak Ridge National Laboratory
McBrayer, James F.	Oak Ridge National Laboratory

PROGRAM AFFILIATION	ACTIVITY
Oak Ridge Site	Land/Water Interaction
	Aquatic Mineral Cycling
	Process Coordinator
Triange Site	Terrestrial Primary Production
Triangle Site	Terrestrial Primary Production
Lake Wingra Site	Aquatic Secondary Production
Lake Wingra Site	Land/Water Interaction
Lake Wingra Site	Hydrology
Lake Wingra Site	Hydrology
Lake George Site	Aquatic Primary Production
Biome and Regional Analysis	
Lake Wingra Site	Land/Water Interaction
	Process Coordinator
Coweeta Site	Terrestrial Decomposition
Oak Ridge Site	Meteorology
Triangle Site	Terrestrial Mineral Cycling
Management	Information Center
Biome and Regional Analysis	
Triangle Site	Terrestrial Mineral Cycling
Management	Advisory Committee
Lake George Site	Hydrology
Lake Wingra Site	Aquatic Mineral Cycling
Management	Modeling
Lake George Site	Aquatic Mineral Cycling
Triangle Site	Terrestrial Primary Production
Lake Wingra Site	Aquatic Secondary Production
Lake Wingra Site	Aquatic Secondary Production
Lake George Site	Aquatic Mineral Cycling
Biome and Regional Analysis	
Lake George Site	Information Center
Lake Wingra Site	Modeling
Triangle Site	Site Coordinator
	Meteorology Process
	Coordinator
Lake George Site	Aquatic Secondary Production
Biome and Regional Analysis	
Biome and Regional Analysis	
Lake Wingra Site	Site Coordinator
Oak Ridge Site	Land/Water Interaction
Oak Ridge Site	Terrestrial Secondary
	Production

NAME	INSTITUTION
McConathy, Ronald K.	Oak Ridge National Laboratory
MacCormick, A. J. A.	University of Wisconsin
McCracken, Michael D.	Texas Christian University
McGinty, Douglas	University of Georgia
McLeod, Robert S.	U.S. Geological Survey
McNaught, Donald C.	State University of New York, Albany
McSwain, Mike	U.S. Forest Service
Madgwick, Herbert A.	Virginia Polytechnic Institute
Magnuson, John J.	University of Wisconsin
Mankin, J. B., Jr.	Oak Ridge National Laboratory
Marks, Peter L.	Cornell University
Matt, Detlaf	Oak Ridge National Laboratory
Metz, Louis J.	U.S. Forest Service
Miller, Edward E.	University of Wisconsin
Mitchell, John	University of Georgia
Monk, Carl D.	University of Georgia
Monkmeyer, Peter L.	University of Wisconsin
Mowry, Fred	Duke University
Murphy, Charles E.	Duke University
Nagy, Robert	Rensselaer Polytechnic Institute
Neuhold, John M.	Utah State University
Noggle, G. Ray	North Carolina State University
Northup, Melvin L.	University of Wisconsin
Olson, Jerry S.	Oak Ridge National Laboratory
O'Neill, Robert V.	Oak Ridge National Laboratory
Park, Richard A.	Rensselaer Polytechnic Institute
Perrotte, William T.	Marist College
Peterson, James	University of Wisconsin
Petty, Robert D.	Wabash College
Pineo, Patricia	Duke University
Ralston, Charles W.	Duke University
Randolph, James C.	Oak Ridge National Laboratory
Reader, Richard	University of North Carolina
Reeves, Mark, III	Oak Ridge National Laboratory
Reichle, David E.	Oak Ridge National Laboratory
Sandwick, Jane	State University of New York, Albany
Schreuder, Hans	U.S. Forest Service
Shanks, Marvin H.	Oak Ridge National Laboratory
Sharpe, David M.	Southern Illinois University

PROGRAM AFFILIATION	ACTIVITY
Oak Ridge Site	Terrestrial Primary Production
Lake Wingra Site	Modeling
Lake Wingra Site	Aquatic Primary Production
Coweeta Site	Terrestrial Primary Production
Lake Wingra Site	Hydrology
Lake George Site	Aquatic Secondary Production Process Coordinator
Coweeta Site	Land/Water Interaction
Management	Advisory Committee
Lake Wingra Site	Aquatic Secondary Production
Management	Modeling
Biome and Regional Analysis	
Oak Ridge Site	Meteorology
Triangle Site	Terrestrial Decomposition
Lake Wingra Site	Land/Water Interaction
Coweeta Site	Information Center
Coweeta Site	Site Coordinator
Lake Wingra Site	Hydrology
Triangle Site	Meteorology
Triangle Site	Modeling
Lake George Site	Information Center
Management	Advisory Committee
Biome and Regional Analysis	
Lake Wingra Site	Terrestrial Mineral Cycling
Management	Foreign Liaison Coordinator
Management	Modeling Coordinator Scientific Director
Lake George Site	Modeling
Lake George Site	Aquatic Secondary Production
Lake Wingra Site	Aquatic Secondary Production
Oak Ridge Site	Terrestrial Primary Production
Triangle Site	Information Center
Triangle Site	Terrestrial Primary Production
Oak Ridge Site	Terrestrial Secondary Production
Biome and Regional Analysis	
Management	Modeling
Oak Ridge Site	Site Coordinator
Lake George Site	Aquatic Secondary Production
Triangle Site	Terrestrial Mineral Cycling
Oak Ridge Site	Terrestrial Mineral Cycling
Biome and Regional Analysis	

NAME	INSTITUTION
Sheppard, John	Oak Ridge National Laboratory
Shetron, Stephen	Michigan Technological Institute
Shugart, Herman H., Jr.	Oak Ridge National Laboratory
Sinclair, Thomas R.	Duke University
Skinner, Earl L.	U.S. Geological Survey
Stearns, C. R.	University of Wisconsin
Stearns, Forest W.	University of Wisconsin, Milwaukee
Sterling, Claudia	Rensselaer Polytechnic Institute
Stephenson, Robert L.	Oak Ridge National Laboratory
Stone, Walter H.	University of Wisconsin
Strain, Boyd R.	Duke University
Strand, Rodney H.	Oak Ridge National Laboratory
Stross, Raymond B.	State University of New York, Albany
Sutherland, James	State University of New York, Albany
Swank, Wayne	U.S. Forest Service
Swindel, Bennie	North Carolina State University
Taylor, Fred G., Jr.	Oak Ridge National Laboratory
Titus, John	University of Wisconsin
Todd, Donald E.	Oak Ridge National Laboratory
Van Hook, Robert I., Jr.	Oak Ridge National Laboratory
Van Winkle, Webster, Jr.	Oak Ridge National Laboratory
Waide, Jack	University of Georgia
Webster, Jack	University of Georgia
Weiler, Peter	University of Wisconsin
Wells, Carol G.	U.S. Forest Service
Wenger, Karl F.	U.S. Forest Service
West, Darrell C.	Oak Ridge National Laboratory
Whittingham, William F.	University of Wisconsin
Wilkinson, John W.	Rensselaer Polytechnic Institute
Williams, Eliot C.	Wabash College
Wuenscher, James	Duke University
Zuehls, Elmer	U.S. Geological Survey

PROGRAM AFFILIATION	ACTIVITY
Oak Ridge Site Biome and Regional Analysis	Hydrology
Oak Ridge Site	Modeling
Triangle Site	Meteorology
Lake Wingra Site	Hydrology
Lake Wingra Site Biome and Regional Analysis	Meteorology
Lake George Site Biome and Regional Analysis	Information Center
Lake Wingra Site	Aquatic Primary Production
Triangle Site	Site Coordinator
Oak Ridge Site	Information Center
Lake George Site	Aquatic Primary Production
Lake George Site	Aquatic Secondary Production
Coweeta Site	Land/Water Interaction
Triangle Site	Terrestrial Mineral Cycling
Oak Ridge Site	Terrestrial Primary Production
Lake Wingra Site	Aquatic Primary Production
Oak Ridge Site	Terrestrial Primary Production
Oak Ridge Site	Terrestrial Secondary Production
Management	Modeling
Coweeta Site	Modeling
Coweeta Site	Modeling
Lake Wingra Site	Information Center
Triangle Site	Terrestrial Mineral Cycling
Management Biome and Regional Analysis	Advisory Committee
Lake Wingra Site	Terrestrial Decomposition
Lake George Site	Modeling
Oak Ridge Site	Terrestrial Secondary Production
Triangle Site	Terrestrial Primary Production
Lake Wingra Site	Hydrology

Coniferous Forest Biome

DIRECTORATE: Director, Stanley P. Gessel
Executive Directorate:
    Burgner, R. L., University of Washington
    Chapman, D. G., University of Washington
    Cole, D. W., University of Washington
    Edmonds, R. L., University of Washington
    Waring, R. H., Oregon State University

RESEARCH SITES:
    Principal Sites:
        Drainage Basin of Cedar River, Cascade Mountains near Seattle, Washington
        H. J. Andrews Experimental Forest (U.S.) Forest Service), Cascade Mountains of Oregon
    Other Sites:
        Cache la Poudre River, Colorado
        Castle Lake, California
        Fairbanks area, Alaska
        Flathead Lake, Montana
        Eagle Lake, California
        Little South Priest Lake, Idaho
        Upper Logan Canyon, Utah
        White Canyon-Cedar Mesa, Utah
        Wood River Lakes, Alaska

DURATION OF PROGRAM: 1969-75

OBJECTIVES: The specific objectives of the program are: (1) To understand the relative behavior of terrestrial and aquatic ecosystems in various environments in the biome. (2) To develop conceptual and computer simulation models that describe nutrient, carbon, and waterflows on a short-term basis (fewer than ten years) and which integrate research results and increase our understanding of coniferous forests and associated aquatic ecosystems. These models are being developed at the process, forest stand or water column, and watershed levels. (3) To develop conceptual and computer simulation models that describe the long-term behavior of coniferous forest ecosystems involving succession and erosion. (4) To understand the linkages between terrestrial and aquatic ecosystems. (5) To determine the effect of manipulations, such as clearcutting, fertilization, and defoliation on terrestrial and aquatic ecosystems.

RESULTS: 296 open literature publications; one volume ("Organization, Behavior, and Response to Stress of Coniferous Forest Ecosystems") in US/IBP series (in preparation).

PARTICIPANTS
Amundson, R., University of Washington
Anderson, A. W., Oregon State University
Anderson, Norman H., Oregon State University
Avery, C., USFS, Flagstaff, Arizona
Bartoo, N., University of Washington
Beck, S., University of Idaho
Behan, Mark J., University of Montana
Belt, George H., University of Idaho
Berggren, T., University of Washington
Birch, P., University of Washington
Bissonnette, P., University of Washington
Black, H., Oregon State University
Bledsoe, Caroline, University of Washington
Bockheim, J., University of Washington
Bradley, B., University of Washington
Brown, Harry E., USDA, Northern Arizona University
Brown, R., Oregon State University
Burges, Stephen J., University of Washington
Burgner, Robert L., University of Washington
Carroll, George C., University of Oregon
Casne, S., University of Washington
Chapman, Douglas G., University of Washington
Cole, Dale W., University of Washington
Cromack, Kermit, U.S. Forest Service, Corvallis, Oregon
Daniel, T. W., Utah State University
Daterman, Gary E., U.S. Forest Service, Corvallis, Oregon
Deitschman, Glenn H., U.S. Forest Service, Moscow, Idaho
DeLacy, Allan C., University of Washington
DelMoral, Roger, University of Washington
Denison, William C., Oregon State University
Devol, A., University of Washington
Deyrup, M., University of Washington
Dirmhirn, Inge, Utah State University
Doble, B., University of Washington
Donaldson, John R., Oregon State University
Doraiswamy, P., University of Washington
Driver, Charles H., III, University of Washington
Dyrness, C. Theodore, U.S. Forest Service, Corvallis, Oregon
Edmonds, Robert L., University of Washington
Eggers, D., University of Washington
Emmingham, William H., Oregon State University
Erickson, C., University of Washington
Farnum, P., University of Washington

Fogel, R., Oregon State University
Forcier, Lawrence K., University of Montana
Fowler, Charles, University of Washington
Franklin, Jerry F., National Science Foundation
Fredriksen, Richard L., U.S. Forest Service, Corvallis, Oregon
Fritschen, Leo J., University of Washington
Froelich, H., Oregon State University
Gara, Robert I., University of Washington
Gaufin, Arden R., University of Utah
Gay, Lloyd W., Oregon State University
Gessel, Stanley P., University of Washington
Gilmour, C. M., University of Idaho
Glenn, L. F., Oregon State University
Goldman, R., University of California, Davis
Gregory, Stan, Oregon State University
Grier, Charles C., University of Washington
Hall, James D., Oregon State University
Hamerly, E. R., University of Washington
Harr, R. Dennis, Oregon State University
Hart, George E., Utah State University
Hatheway, William H., University of Washington
Hawk, Glen, U.S. Forest Service, Corvallis, Oregon
Hawkins, Richard H., Utah State University
Haydu, E. P., Weyerhauser, Longview, Washington
Helms, John A., University of California, Berkeley
Henderson, C., Utah State University
Hendry, G., University of Washington
Hermann, Richard K., Oregon State University
Hett, Joan M., University of Washington
Hill, R., Utah State University
Holbo, R., University of Washington
James, M., University of Oregon
Jensen, Harold J., Oregon State University
Johnson, Dale, University of Washington
Johnson, Fred D., University of Idaho
Kays, M. Allan, University of Oregon
Kickert, Ron N., University of California
Kittle, L., University of Washington
Kline, Jerry L., Argonne National Laboratory
Krantz, Gerald W., Oregon State University
Kurmes, Ernest A., Northern Arizona University
Lassoie, James, Cornell University
Lavender, Denis P., Oregon State University
Leverenz, J., University of Washington

Lighthart, Bruce, Environmental Protection Agency, Corvallis, Oregon
Long, James, University of Washington
Lyford, John H., Jr., Oregon State University
Machno, P., University of Washington
Male, L., University of Washington
Malick, J., University of Washington
Matches, Jack R., University of Washington
McIntire, David C., Oregon State University
McKee, Art, U.S. Forest Service, Corvallis, Oregon
Miller, S., University of Washington
Minden, R., University of Washington
Minyard, P., University of Washington
Mogren, Edwin W., Colorado State University
Moir, W., Colorado State University
Monahan, F., University of Washington
Moore, Duane G., U.S. Forest Service, Corvallis, Oregon
Nagel, William P., Oregon State University
Nellis, C., University of Washington
Nussbaum, Ronald A., Oregon State University
Olson, Paul R., City of Seattle Water Department
Overton, W. Scott, Oregon State University
Packard, Theodore T., University of Washington
Pamatmat, M. M., Auburn University
Parsons, R., Oregon State University
Paulsen, Dennis R., University of Washington
Perkins, J., University of Oregon
Pike, Lawrence H., George Mason University
Rains, D. W., University of California, Davis
Rau, G., University of Washington
Reed, Kenneth L., Yale University
Reid, Charles P., Colorado State University
Rhoades, F., Oregon State University
Richey, Jeff, University of Washington
Richter, K., University of Washington
Riekerk, Hans, University of Washington
Riley, J. Paul, Utah State University
Rydell, Rob, U.S. Forest Service, Corvallis, Oregon
Salo, D. J., University of Washington
Santantonio, D., Oregon State University
Schoen, J., University of Washington
Scott, D. R. M., University of Washington
Sedell, James R., Oregon State University
Sherk, R., University of Washington
Sherwood, M., Oregon State University

Shih, G. D., Utah State University
Singer, Michael J., University of California, Davis
Smith, S., University of Washington
Smith, M., University of Washington
Sollins, Phillip, Oregon State University
Spyridakis, Demetrios E., University of Washington
Staley, James T., University of Washington
Steinhoff, Harold W., Colorado State University
Stettler, Reinhard F., University of Washington
Stober, Quentin J., University of Washington
Strand, Mary Ann, Boyce Thompson Institute, New York
Swanson, F. J., University of Oregon
Swanston, D. N., U.S. Forest Service, Corvallis, Oregon
Swartzman, Gordon, University of Washington
Taber, Richard D., University of Washington
Taub, Frieda B., University of Washington
Taylor, Alan R., University of Montana
Terry, C., University of Washington
Thompson, Graham, University of Montana
Thorne, Richard E., University of Washington
Tracy, D., Oregon State University
Trappe, J., U.S. Forest Service, Corvallis, Oregon
Traynor, J., University of Washington
Triska, Frank, Oregon State University
Tsukada, Matsuo, University of Washington
Turner, J., University of Washington
Ugolini, Florenzo C., University of Washington
Van Cleve, Keith, University of Alaska
Walker, Richard B., University of Washington
Wamback, Robert F., University of Montana
Waring, Richard H., Oregon State University
Warren, Charles E., Oregon State University
Webb, Warren L., Oregon State University
Wekell, M., University of Washington
Welch, Eugene B., University of Washington
Wellner, Charles A., U.S. Forest Service, Moscow, Idaho
Wernz, J., Oregon State University
White, Curtis, Oregon State University
Whitney, Richard R., University of Washington
Wissmar, Robert, University of Washington
Wydoski, Richard S., Utah State University
Wyman, K., University of Washington
Youngberg, Chester T., Oregon State University
Zasoski, R. J., University of Washington
Zobel, Donald B., Oregon State University

Desert Biome

DIRECTORATE: Director, David W. Goodall, 1968-71; Co-Directors, David W. Goodall and Frederic H. Wagner, 1971-73; Director, Frederic H. Wagner, 1973-76.

DURATION OF PROGRAM: 1969-76

RESEARCH SITES:
Principal Sites
    Curlew Valley, Utah
    Pine Valley, Utah
    Rock Valley, Nevada
    Silverbell Bajada, Arizona
    Jornada, New Mexico
Other Sites
    Deep Canyon, California
    Saratoga Springs, California
    Sycamore Creek, Arizona

OBJECTIVES: To elaborate the structure and function of the desert ecosystem. The operational goal was the development of computer simulation models of the system which would represent the numerous simultaneous and interactive functions taking place in the system and to attempt to validate these models.

RESULTS: Fifty-five papers in open literature (to June 1973), three volumes in preparation for US/IBP Series ("Nitrogen Processes in Deserts," "Structure and Function of Desert Ecosystems," and "The Hydrologic Cycle in Desert Ecosystems").

PARTICIPANTS
Soil Water and Hydrology
Blincoe, C., University of Nevada
Cable, Dwight R., U.S. Forest Service, Tucson, Arizona
Campbell, Gaylon S., Washington State University
Carpelen, L., University of California, Riverside
Colthrap, George B., Utah State University
Evans, Daniel D., University of Arizona
Gifford, Gerald F., Utah State University
Hanks, R. John, Utah State University
Harris, Grant A., Washington State University
Hart, George E., Jr., Utah State University
Letey, John, Jr., University of California, Riverside
Mehuys, G. R., University of California, Riverside
Ondrechen, W. T., Utah State University

Qashu, Hasan K., University of Arizona
Ryan, J., University of Arizona
Sammis, Theodore, University of Arizona
Schmutz, Erwin M., University of Arizona
Stolzy, Lewis H., University of California, Riverside
Thames, John L., University of Arizona
Walk, Fred, Jr., Utah State University
Weeks, L. V., University of California, Riverside
Wheeler, Merlin L., University of Arizona

Climatology and Micrometeorology
Dirmhirn, Inge, Utah State University
Eaton, F., Utah State University
Kaaz, Howard W., University of California, Los Angeles
Thames, John L., University of Arizona

Terrestrial Nutrient Cycling
Ashcroft, Gaylen L., Utah State University
Childress, J., Mercury, Nevada
Dutt, Gordon R., University of Arizona
Eberhardt, P., University of Arizona
Farnsworth, Raymond B., Brigham Young University
Fletcher, Joel E., Utah State University
Fuller, Wallace H., University of Arizona
Griffin, R. A., University of Illinois
Hadfield, K. L., Brigham Young University
Hendricks, D. M., University of Arizona
James, David W., Utah State University
Jurinak, Jerome J., Utah State University
Kline, Larry G., Utah State University
Kotter, Sally, Utah State University
MacGregor, A. N., University of Arizona
McKell, Cyrus M., Utah State University
Porcella, Donald B., Utah State University
Romney, Evan M., University of California, Los Angeles
Skujins, John J., Utah State University
Southard, Alvin R., Utah State University
Staffeldt, Eugene E., New Mexico State University
Stark, Nellie M. B., Montana State University
Tucker, Thomas C., University of Arizona
Utter, G., Utah State University
Wallace, Arthur, University of California, Los Angeles
West, Neil E., Utah State University
Westerman, Robert L., University of Arizona
Wilson, Lemoyne, Utah State University

Aquatic Nutrient Cycling
Post, Frederick J., Utah State University
Walkiw, Irene, Utah State University

Terrestrial Primary Production
Ackerman, T., University of California, Los Angeles
Ashcroft, R., University of California, Los Angeles
Balda, Russell P., Northern Arizona University
Balding, Frederick R., New Mexico State University
Bamberg, Samuel A., University of California, Los Angeles
Bartlett, E. Thomas, University of Arizona
Beatley, Janice C., University of Cincinnati
Braun, C., Utah State University
Brewster, Sam F., Jr., Ricks College, Idaho
Caldwell, Martyn M., Utah State University
Cameron, Roy E., California Institute of Technology, Pasadena
Childs, S., Utah State University
Cunningham, Gary L., New Mexico State University
De Puit, E. J., Utah State University
Fareed, Marci, Utah State University
Faust, W., University of Arizona
Fernandez, O., Utah State University
Fish, E., University of Arizona
Franz, C. E., Northern Arizona University
Goodall, David W., Utah State University
Goodman, Peter J., Welsh Plant Breeding Station, Aberystwyth, Wales
Hale, Verle Q., University of California, Los Angeles
Hironaka, M., University of Idaho
Holte, Karl E., Idaho State University
Johnson, Hyrum B., University of California, Riverside
Kleinkopf, G., University of California, Los Angeles
Klemmedson, James O., University of Arizona
Klikoff, Lionel G., University of Utah
Lindsay, John H., University of Arizona
Ludwig, John, New Mexico State University
Lynn, Raymond I., Utah State University
Marsh, Gordon, University of California, Irvine
Moore, T. R., Utah State University
Nash, Thomas H., III, Arizona State University
Nickell, G., New Mexico State University
Nisbet, R. A., Arizona State University
Norton, Brien E., Utah State University
Patten, Duncan T., Arizona State University
Pearson, Lorenz C., Ricks College, Idaho

Reynolds, J., New Mexico State University
Shinn, R., Utah State University
Smith, E. L., University of Arizona
Szarek, S. R., University of California
Ting, Irwin P., University of California, Riverside
Tisdale, E. W., University of Idaho
Tueller, Paul T., University of Nevada
Vollmer, A., University of California, Los Angeles
Wallace, Arthur, University of California, Los Angeles
West, Neil E., Utah State University
White, R. S., Utah State University
Whitson, Paul D., University of Northern Iowa
Wiebe, Herman H., Utah State University

Aquatic Primary Production
Bradley, W. Glen, University of Nevada
Deacon, James E., University of Nevada
Gordon, Robert W., Texas Tech University
Minkley, Wayne L., Arizona State University
Rose, Fred L., Idaho State University

Aquatic Consumers
Ashdown, D., Texas Tech University
Baepler, Donald H., University of Nevada
Bjornn, Theodore C., University of Idaho
Cheng, F. C., University of Idaho
Deacon, James E., University of Nevada
Gaufin, Arden R., University of Utah
Hallmark, M. D., Texas Tech University
Huddleston, E. W., Texas Tech University
Holman, Josephine A., University of Calgary
Dramer, R. H., Utah State University
Meenahan, George F., Texas Tech University
Merritt, M. D., Texas Tech University
Minshall, G. Wayne, Idaho State University
Rajagopal, P. K., Utah State University
Richardson, L. G., Texas Tech University
Wallace, Richard L., University of Idaho
Ward, Charles R., Texas Tech University

Terrestrial Secondary Production
Anderson, Robert D., Idaho State University
Aschwander, A., Atomic Energy Commission, Mercury, Nevada
Austin, G., University of Arizona
Balph, David F., Utah State University

Bateman, Gary C., Northern Arizona University
Bender, Gordon L., Arizona State University
Brown, James Hemphill, University of Utah
Brusven, Merlyn A., University of Idaho
Chew, Robert, University of Southern California
Cockrum, E. Lendell, University of Arizona
Costa, Wayne R., University of California, Riverside
Crawford, Clifford S., University of New Mexico
Dingman, Ross E., University of San Diego
Dixon, Keith L., Utah State University
Edney, Eric B., University of California, Los Angeles
Ettershank, G., Monash University, Australia
Freckman, Diana, University of California, Riverside
Gould, P., University of Arizona
Gray, J. P., University of Arizona
Hanson, Wilford J., Utah State University
Haverty, Michael I., University of Arizona
Hoagstrom, C., University of Arizona
Hsiao, Ting H., Utah State University
Hungerford, Charles R., University of Arizona
Johnson, Clarence D., Northern Arizona University
Johnson, Donald R., University of Idaho
Jorgensen, Clive D., Brigham Young University
Knowlton, George F., Utah State University
LaFage, Jeffrey P., University of Arizona
Lowe, Charles H., University of Arizona
Lowrie, Donald C., California State College, Los Angeles
MacMahon, James, Utah State University
Malechek, John C., Utah State University
Mankau, Reinhold, University of California, Riverside
Maza, Bernardo, University of California, Los Angeles
McBrayer, James F., University of California, Los Angeles
Medica, Philip A., University of California, Los Angeles
Mispagel, Michael, California State University, Long Beach
Murray, S. L., University of Arizona
Nagy, K. A., University of California, Riverside
Nelson, J., University of Southern California
Nutting, William L., University of Arizona
Patten, Duncan T., Arizona State University
Pimm, Stuart, New Mexico State University
Raitt, Ralph J., New Mexico State University
Reichman, James, University of Utah
Reynolds, Hudson, Arizona State University
Russell, Stephen M., University of Arizona

Saul, William E., Idaho State University
Schreiber, R. Kent, Oak Ridge National Laboratory
Scott, D. T., Brigham Young University
Sher, S. A., University of California, Riverside
Sferra, P. R., College of Mount St. Joseph-on-the-Ohio, Ohio
Shoemaker, Vaughan, University of California, Riverside
Sleeper, Elbert L., California State University, Long Beach
Smith, H. Duane, Brigham Young University
Soholt, Lars, Desert Research Institute, Boulder City, Nevada
Stockton, W. D., California State University, Long Beach
Stoddart, L. Charles, Utah State University
Tevis, Lloyd, University of California, Riverside
Trost, C. H., Idaho State University
Turkowski, F., Arizona State University
Turner, Frederick B., University of California, Los Angeles
Van de Graaff, Kent M., Northern Arizona University
Van Gundy, Seymour D., University of California, Riverside
Vaughan, Terry A., Northern Arizona University
Wagner, Frederic H., Utah State University
Werner, Floyd G., University of Arizona
Whitford, Walter G., New Mexico State University
Wirtz, Willard O., II, Pomona College, California
Zimmerman, James R., New Mexico State University

Resource Management Modeling
Norton, Brien E., Utah State University
Romesberg, Charles, Utah State University
Wilkin, Donovan C., Utah State University

Terrestrial Decomposition and Mineralization
Au, Fred H., U.S. Public Health Service, Las Vegas, Nevada
Comanor, Peter L., University of Nevada
Fareed, Marci, Utah State University
Loebeck, Maude E., Idaho State University
O'Brien, Robert T., New Mexico State University
Prusso, Don C., University of Nevada
West, Neil E., Utah State University
Wollum, Arthur G., New Mexico State University

Terrestrial Ecosystem Modeling
Bridges, Kent W., University of Hawaii
Crawley, Michael, Utah State University
Gilbert, Bradley, Utah State University
Gist, Clayton, Utah State University

Goodall, David W., Utah State University
Griffin, R., University of Illinois
Heasley, John, Utah State University
Innis, George S., Utah State University
Kickert, Ronald, Utah State University
Lommen, Paul, Utah State University
Marshall, Kim, Utah State University
Olsen, Alma D., Utah State University
Parnas, Hanna, Hebrew University, Israel
Payne, Susan, Utah State University
Radford, John, Utah State University
Reynolds, James, New Mexico State University
Schultze, E. D., University of Wurzburg, Germany
Valentine, Walter, Utah State University
Watson, James D., Utah State University
Westoby, Mark, Utah State University
Wilcott, Curtis, Utah State University

Aquatic Ecosystem Modeling
Goodall, David W., Utah State University
Greeney, W., Utah State University
Holman, Josephine A., University of Calgary, Canada
Huber, A. Leon, Utah State University
Koob, Derry, Utah State University
Minshall, G. Wayne, Idaho State University
Minshall, J. N., Idaho State University
Porcella, Donald B., Utah State University
Stalnaker, Clair B., Utah State University
Wlosinski, Joseph, Utah State University

Tundra Biome

PROGRAM DIRECTORATE: Director, Jerry Brown; Director Tundra Biome Center, George C. West.

RESEARCH SITES: Barrow (intensive site), Prudhoe Bay, Eagle Summit, Alaska; and Niwot Ridge, Colorado.

DURATION OF FIELD PROGRAM: 1970-74

DURATION OF SYNTHESIS PROGRAM: 1975-76

OBJECTIVES: To analyze the structure and function of the wet arctic tundra so as to develop a predictive understanding of this and related tundra ecosystems.

RESULTS: 150 papers in open literature (to October 1976); two volumes in US/IBP Synthesis Series (Aquatic and Terrestrial, in preparation): one volume on Prudhoe Bay; one volume producers; and major contribution to three international tundra volumes (Sweden Ecol. Bull. 20 (1975), Soil Organisms and Decomposition in Tundra (1974), and CUP (in prep.)); 19 Ph.D. and 26 Master's theses (to October 1976).

SENIOR PARTICIPANTS
Primary Production
Allessio (Leck.), Mary, Rider College, New Jersey
Anderson, James H., University of Alaska
Billings, W. Dwight, Duke University, North Carolina
Caldwell, Martyn M., Utah State University
Chapin, F. Stuart, III, University of Alaska
Collins, Nigil, British Antarctic Survey
Dennis, John G., National Park Service
Johnson, A. W., San Diego State University
Koranda, John, Lawrence Radiation Laboratory
Lewis, Martin, University of Toronto, Canada
McCown, Brent H., University of Wisconsin
Miller, Philip C., San Diego State University
Murray, David F., University of Alaska
Nieland, Bonita J., University of Alaska
Oechel, Walter C., McGill University, Montreal, Canada
Packer, John, University of Alberta, Canada
Rastorfer, James R., Chicago State University
Rudolph, Emanuel D., Ohio State University
Schofield, Ed, Ohio State University
Tieszen, Larry L., Augustana College, Illinois
Ulrich, Albert, University of California, Berkeley
Webber, Patrick J., University of Colorado

Applied Aspects of Plant Growth
Bonde, Eric K., University of Colorado
Foreman, Maxine F., Community College, Denver, Colorado
Deneke, Frederick, Kansas State University
McCown, Brent H., University of Wisconsin
McKendrick, Jay D., University of Alaska
Mitchell, William W., University of Alaska
Rickard, Warren, no present affiliation
Wooding, Frank J., University of Alaska

Small Mammalian Consumers
Banks, Edwin M., University of Illinois
Batzli, George O., University of Illinois
Feist, Dale D., University of Alaska
Fitzgerald, B. Michael, DSIR, New Zealand
MacLean, Stephen F., Jr., University of Alaska
Melchior, Herbert R., University of Alaska
Pitelka, Frank A., University of California, Berkeley

Large Mammalian Consumers
Dieterich, Robert A., University of Alaska
Henshaw, Robert E., University of Pennsylvania
Holleman, Dan F., University of Alaska
Klein, David R., University of Alaska
Luick, Jack R., University of Alaska
Skogland, Terje, Statens Viltundersokelser, Norway
Thomson, Brian, University of Edinburgh, Scotland
Underwood, Larry S., University of Alaska at Anchorage
White, Robert G., University of Alaska

Avian Consumers
Braun, Clait E., Colorado Division of Game, Fish, and Parks
Cade, Tom J., Cornell University, New York
DeWolfe, Barbara B., University of California, Santa Barbara
Haugh, John R., Energy Research Development Administration
MacLean, Stephen F., Jr., University of Alaska
Norton, David W., University of Alaska
Pitelka, Frank A., University of California, Berkeley
Safriel, Uriel N., Hebrew University, Israel
Sladen, William J. L., Johns Hopkins University, Maryland
West, George C., University of Alaska
White, Clayton M., Brigham Young University, Utah

Invertebrate Consumers
Collier, Boyd D., San Diego State University
Crossley, D. A., University of Georgia
Edwards, John S., University of Washington
MacLean, Stephen F., Jr., University of Alaska
Philip, Kenelm W., University of Alaska
Schmoller, Ronald, University of Tennessee
Smart, Grover C., Jr., University of Florida

Decomposers, Nutrient Flux, and Soils
Alexander, Vera, University of Alaska
Arkley, Rodney J., University of California, Berkeley
Atlas, Ronald, University of Louisville, Kentucky
Barel, D., Iowa State University
Barsdate, Robert J., University of Alaska
Benoit, Robert E., Virginia Polytechnic Institute
Cameron, Roy E., Argonne National Laboratory, Illinois
Douglas, Lowell A., Rutgers University, New Jersey
Everett, Kaye R., Ohio State University
Flanagan, Patrick, University of Alaska
Flint, Philip, University of California, Berkeley
Gersper, Paul L., University of California, Berkeley
Glaser, Rudi, University of California, Berkeley
Kallio, S., University of Turku, Finland
Miller, Orson K., Virginia Polytechnic Institute
Norrell, Stephen A., University of Alaska
Scarborough, Arla, University of Alaska
Schell, Donald M., University of Alaska
Schultz, Arnold, University of California, Berkeley
Van Cleve, Keith, University of Alaska

Abiotic (Climate and Meteorological)
Barry, Roger G., University of Colorado
Benson, Carl S., University of Alaska
Berglund, Erwin R., University of Alaska
Brown, Jerry, USA CRREL
Coyne, Patrick, II, Lawrence Radiation Laboratory
Dingman, S. Lawrence, University of New Hampshire
Guymon, Gary, University of California, Irvine
Haugen, Richard, USA CRREL
Holmgren, Bjorn, Meteorolngiska Institute, Sweden
Ives, Jack D., University of Colorado
Kelley, John J., Sr., University of Alaska
Lewellen, Robert, Arctic Consultant, Littleton, Colorado

Lord, Norman W., Center of Environment and Man
Nakano, Yoshisuke, USA CRREL
Pandolfo, Joseph P., Center of Environment and Man
Parrish, Scott, University of Alaska (deceased)
Sykes, Dwane, University of Alaska
Weller, Gunter, University of Alaska

Aquatic
Alexander, Vera, University of Alaska
Barsdate, Robert J., University of Alaska
Bierle, Donald A., Sioux Falls Bible School
Cameron, James N., University of Alaska
Dillon, Raymond D., University of South Dakota
Dodson, Stanley, University of Wisconsin
Fenchel, Tom, University of Aarhus, Denmark
Hobbie, John E., Woods Hole Marine Biology Laboratory
Kangas, Donald, Northeast Missouri State University
McRoy, C. Peter, University of Alaska
Miller, Michael C., University of Cincinnati, Ohio
Stanley, Donald, North Carolina State University
Stross, Raymond G., State University of New York, Albany
Tiwari, Jawahar, Woods Hole Marine Biology Laboratory

Modeling
Bridges, K. W., University of Hawaii
Brittain, E. G., Australian National University, Canberra
Bunnell, Fred L., University of British Columbia, Canada
Collier, Boyd D., San Diego State University
Coulombe, H. N., San Diego State University
Dean, Frederick C., University of Alaska
Dingman, Lawrence, University of New Hampshire
Harbo, Samuel J., University of Alaska
Miller, Philip C., San Diego State University
Murray, Jere, no present affiliation
Nakano, Yoshisuke, USA CRREL
Outcalt, Samuel, University of Michigan
Timin, Mitchell, no present affiliation
Tiwari, Jawahar, Woods Hole Marine Biology Laboratory

ORIGIN AND STRUCTURE OF ECOSYSTEMS

DIRECTORATE: Director, W. Frank Blair

OBJECTIVES: To integrate studies of ecology with studies of evolution to investigate ecosystem structure and development.

Desert Scrub Project

DIRECTORATE: Director, Otto T. Solbrig

PRINCIPAL RESEARCH SITES: Silver Bell Bajada, near Tucson, Arizona, USA; Bolsón de Pipanaco, near Andalgalá, Catamarca, Argentina

DURATION OF PROGRAM: 1969-75

OBJECTIVES: To compare the structure and function of desert scrub ecosystems (Sonoran Desert and Argentine Monte Desert) that have relatively similar physical environments. To determine the degree of similarity of these ecosystems and to learn how much of the similarity is attributable to parallel evolution of closely related taxa and how much to convergence of distantly related taxa.

RESULTS: 39 papers in open literature (to June 1973). Three volumes in US/IBP Synthesis Series (in press).

PARTICIPANTS
Bailey, Harry P., University of California, Riverside
Barbour, Michael G., University of California, Davis
Bawa, Kamaljit, Harvard University, Massachusetts
Birkhead, William S., University of Texas, Austin
Blair, W. Frank, University of Texas, Austin
Bogart, James P., University of Texas, Austin
Bohnstedt, Charles, University of Texas, Austin
Brown, Gordon, Harvard University, Massachusetts
Cantino, Phillip, Harvard University, Massachusetts
Carman, Neil, University of Texas, Austin
Cates, Rex G., University of New Mexico
Cross, John, University of Arizona
Diaz, David, University of California, Davis
DiFeo, Dan, University of Texas, Austin
di Tada, Ismael, Universidad de Córdoba, Argentina
Enders, Frank, University of Texas, Austin

Goldstein, Guillermo, INTA, Centro Nacional de Investigaciones Agropecuarias, Buenos Aires

Greegor, David, University of Arizona

Hulse, Arthur C., University of Texas, Austin

Hunsaker, Don, II, San Diego State University

Hunziker, Juan, Universidad de Buenos Aires, Argentina

Hurd, Paul D., Jr., Smithsonian Institution, Washington, D.C.

Joern, Anthony, University of Texas, Austin

Kingsolver, John M., ARS, USDA, Washington, D.C.

LeClaire, Jerry, Harvard University, Massachusetts

Linsley, E. Gordon, University of California, Berkeley

Lowe, Charles H., University of Arizona

Mabry, Tom J., University of Texas, Austin

Mares, Michael A., University of Pittsburgh, Pennsylvania

Moldenke, Andrew, University of California, Santa Cruz

Naranjo, Carlos A., Universidad de Buenos Aires, Argentina

Neff, Jack, University of California, Santa Cruz

Orians, Gordon H., University of Washington

Otte, Daniel, Philadelphia Academy of Sciences, Pennsylvania

Palacios, Ramon A., Universidad de Buenos Aires, Argentina

Poggio, Lidia, Universidad de Buenos Aires, Argentina

Reppun, Paul, Harvard University, Massachusetts

Rhoades, David F., University of Washington

Rodriguez, Eloy, University of Texas, Austin

Rosenzweig, Michael L., University of Arizona

Ross, Herbert H., University of Georgia

Roughgarden, Jonathan, Stanford University, California

Sage, Richard D., University of California, Berkeley

Sanderson, Stewart, University of Texas, Austin

Schultz, John C., Dartmouth College, New Hampshire

Selander, R. B., University of Illinois

Simpson, Beryl B., Smithsonian Institution, Washington, D.C.

Stange, Lionel, Instituto Miguel Lillo, San Miguel de Tucumán, Argentina

Teran, Arturo, Instituto Miguel Lillo, San Miguel de Tucumán, Argentina

Timmerman, Barbara, University of Texas, Austin

Tomoff, Carl S., University of Arizona

Turner, B. L., University of Texas, Austin

Valesi, Amalia, Universidad de Buenos Aires, Argentina

Vander Velde, George, University of Texas, Austin

Vervoorst, Federico, Instituto Miguel Lillo, San Miguel de Tucumán, Argentina

Yang, Tien Wei, University of Arizona

Mediterranean Scrub Project

DIRECTORATE: Co-Directors, Harold A. Mooney (USA), and Ernst R. Hajek (Chile)

PRINCIPAL RESEARCH SITES: Echo Valley, near San Diego, California, and Cuesta la Dormida, Province of Santiago, Chile

DURATION OF PROGRAM: 1969-75

OBJECTIVES: Same as for Desert Scrub Project, but between California chaparral and Chilean matorral, and additionally to make microclimatic comparisons between the two.

RESULTS: 30 publications in open literature (to June 1973), one book published (Springer-Verlag 1974), two books in US/IBP Synthesis Series (in preparation).

PARTICIPANTS

Albright, David, San Diego State University
Aljaro, Maria-Ester, Universidad Catolica de Chile, Santiago
Araya, Sandra, Universidad Catolica de Chile, Santiago
Aschmann, Homer, University of California, Riverside
Atkins, Michael D., San Diego State University
Avendano, S. Vilma, Universidad Catolica de Chile, Valparaiso
Avila, Guacolda, Universidad Catolica de Chile, Santiago
Bahre, Conrad J., University of California, Riverside
Beauchamp, Mitch, San Diego State University
Bebout, Roberta, San Diego State University
Becking, Lucille Wesley, San Diego State University
Boggs, Jerry, San Diego State University
Botten, Ruth, San Diego State University
Bradbury, David E., University of California, Los Angeles
Brizzolara, Shawna, San Diego State University
Carpenter, F. Lynn, University of California, Berkeley
Castonguay, Tom, San Diego State University
Chu, Celia, Stanford University, California
Cody, Martin L., University of California, Los Angeles
Colwell, Robert K., University of California, Berkeley
Culberson, Chicita, Duke University, North Carolina
Culberson, William L., Duke University, North Carolina
Damm, Albert, Universidad Catolica de Chile, Santiago
de la Puente, Fernando Riveros, Universidad Catolica de Chile, Santiago
Dement, Karen, Stanford University, California
Dement, William, Stanford University, California
DeRemer, Sherry, Stanford University, California

Diaz, Noel, University of California, Los Angeles
di Castri, Francesco, Universidad Austral de Chile, Valdivia
di Castri, Vitali, Universidad Austral de Chile, Valdivia
DiFeo, Dan, University of Texas, Austin
Dodge, J. Marvin, University of California, Riverside
Duffy, Mary, San Diego State University
Ehleringer, Edna, San Diego State University
Enders, Frank, University of Texas, Austin
Estay, Hiram, Universidad Catolica de Chile, Santiago
Fishbeck, Kathleen, San Diego State University
Fuentes, Eduardo, University of California, Berkeley
Gabbard, Kenneth, Duke University, North Carolina
Garrison, Susan, Stanford University, California
Gemeroy, Sally, San Diego State University
Giliberto, Juan, Universidad Catolica de Chile, Santiago
Glanz, William, University of California, Berkeley
Gonzales, Louis, Universidad Catolica de Chile, Santiago
Gulmon, Sherry, Stanford University, California
Hajek, Ernesto, Universidad Catolica de Chile, Santiago
Hays, Rachel I., San Diego State University
Hays, Robert L., San Diego State University
Hoffmann, Adriana, Universidad Catolica de Chile, Santiago
Hoffmann, Alicia, Universidad Catolica de Chile, Santiago
Hunt, James H., University of California, Berkeley
Hurtubia, Jaime, Universidad Austral de Chile, Valdivia
Hutchison, Steve, San Diego State University
Hynum, Barry, San Diego State University
Johnson, Albert W., San Diego State University
Johnson, Barbara, San Diego State University
Johnson, Chris, San Diego State University
Johnson, Susan, San Diego State University
Jow, Bill, San Diego State University
Keeley, Jon, San Diego State University
Keeley, Sterling, University of Georgia
Krause, David, San Diego State University
Kummerow, Jochen, Universidad Catolica de Chile, Santiago
LaMarche, Valmore, University of Arizona
Lawrence, William, San Diego State University
LeBoulenge, Paule, Universidad Catolica de Chile, Santiago
Lee, Janet, San Diego State University
Lilley, Barbara, Stanford University, California
Lincoln, Patricia, University of California, Santa Cruz
Mabry, Tom, University of Texas, Austin
Martinez, Emiliana, Universidad Catolica de Chile, Santiago

McBarron, Philip, San Diego State University
McClanahan, Jackie, San Diego State University
Miller, Patricia, San Diego State University
Miller, Philip C., San Diego State University
Moldenke, Andrew R., University of California, Santa Cruz
Molina, Juan Domingo, Universidad Catolica de Chile, Santiago
Montenegro, Gloria, Universidad Catolica de Chile, Santiago
Mooney, Harold, Stanford University, California
Moore, Russel T., San Diego State University
Neff, John, University of California, Santa Cruz
Ng, Edward, San Diego State University
Parsons, David, Stanford University, California
Pincetl, Pierre, San Diego State University
Pincetl, Stephanie, San Diego State University
Pitelka, Sandra, Stanford University, California
Poole, Dennis, San Diego State University
Rafols, Eugenio Sierra, Universidad Catolica de Chile, Santiago
Raven, Peter H., Missouri Botanical Garden
Rieden, Chris, San Diego State University
Robbins, Chris, San Diego State University
Roberts, Steve, San Diego State University
Roberts, Eda, San Diego State University
Rogers, Jeff, San Diego State University
Roveraro, Carlos, Universidad Catolica de Chile, Santiago
Saiz, Francisco, Universidad Catolica de Chile, Valparaiso
Salgado, Mercedes, Universidad Catolica de Chile, Santiago
Sanchez, Eduardo L., Catedra de Quimica, Santiago, Chile
Seguy, Andres, Universidad Catolica de Chile, Santiago
Senger, Leslie W., University of California, Los Angeles
Slade, Norman, San Diego State University
Smith, William K., San Diego State University
Solbrig, Otto, Harvard University, Massachusetts
Sutherland, Marsha Pohl, San Diego State University
Thompson, Wallace B., University of Texas, Austin
Thrower, Norman J. W., University of California, Los Angeles
Toro, Haroldo, Universidad Catolica de Chile, Valparaiso
Tremper, Ronald, San Diego State University
Van Curen, Richard, University of California, Riverside
Vasquez, Enrique M., Universidad Catolica de Chile, Valparaiso
Westfall, Mary Ann, San Diego State University
Woodward, Susan, University of California, Los Angeles
Wosika, Ed, San Diego State University

Persons Serving as Project Advisers
Axelrod, Daniel
Bailey, Harry
Blair, W. Frank
Janzen, Daniel
Loucks, Orie
Orme, Anthony
Pratt, P.
Raven, Peter
Roughgarden, Jonathan
Sauer, Jonathan

Persons Assisting in Taxonomic Identifications
Arnaud, Paul
Bohart, George
Bohart, Richard
Daly, Howell
Downie, Norman
Eickwort, George
Garaventa, Agustin
Grigarick, Albert
Hall, Jack
Hurd, Paul
La Berge, Wallace
Leech, Hugh
Michener, Charles
Pena, Louis
Ricardi , M. Mario
Rojas, Fresia
Rozen, Jerome, Jr.
Rust, Richard
Schlinger, Evert
Snelling, Roy
Thorp, Robbin
Timberlake, P. H.
Vockeroth, J. Richard
Zöllner, Otto

Island Ecosystem Stability and Evolution

DIRECTORATE: Director, Dieter Mueller-Dombois; Co-Directors, Andrew J. Berger and J. Linsley Gressitt; Assistant Director, Kent W. Bridges

PRINCIPAL RESEARCH SITE: Hawaii Volcanoes National Park, Island of Hawaii

DURATION OF PROGRAM: 1971-75

OBJECTIVES: To study the biological organization of selected Hawaiian ecosystems and to evaluate the interrelationships of native and exotic species.

RESULTS: 67 papers in open literature (to October 1976), one book in US/IBP Synthesis Series (in preparation).

PARTICIPANTS
Araki, Lynnette, University of Hawaii
Ashton, Geoffrey C., University of Hawaii
Baker, Gladys E., University of Hawaii
Balakrishnan, Nadarajah, University of Hawaii
Beardsley, John W., University of Hawaii
Becker, Richard E., University of Hawaii
Bianchi, Fred, Bernice P. Bishop Museum, Hawaii
Brennan, Barry M., University of Hawaii
Bridges, Kent W., University of Hawaii
Carey, G. Virginia, University of Hawaii
Carpenter, F. Lynn, University of California, Irvine
Carson, Hampton L., University of Hawaii
Carson, Johnny, University of North Carolina
Conant, Michael, University of Hawaii
Conant, Sheila, University of Hawaii
Cooray, Ranjit G., University of Hawaii
Corn, Carolyn A., University of Hawaii
Cuddihy, Linda J., University of Hawaii
Davis, Clifton J., State Department of Agriculture, Hawaii
Delfinado, Mercedes D., New York State Museum and Science Service
Doty, Maxwell S., University of Hawaii
Dunn, Paul H., California State University, Fullerton
Ekern, Paul C., University of Hawaii
Friend, Douglas J., University of Hawaii
Fujii, Douglas T., University of Hawaii
Gagné, Wayne C., Bernice P. Bishop Museum, Hawaii
Gay, Ruth A., University of Hawaii

Goff, M. Lee, Bernice P. Bishop Museum, Hawaii
Guest, Sandra, University of Hawaii
Haramoto, Frank H., University of Hawaii
Hardy, D. Elmo, University of Hawaii
Hiesey, William M., Carnegie Institution of Washington, Stanford
Higa, Carol, Bernice P. Bishop Museum, Hawaii
Hirai, Lawrence T., University of Hawaii
Howarth, Francis K., Bernice P. Bishop Museum, Hawaii
Jacobi, James D., University of Hawaii
Johnson, Walter E., Case Western Reserve University, Ohio
Juvik, James O., University of Hawaii, Hilo
Kambysellis, E. Craddock, New York University
Kambysellis, Michael P., New York University
Kamil, Alan C., University of Massachusetts
Kaneshiro, Kenneth Y., University of Hawaii
Kjargaard, John, National Park Service, Hawaii
Lamoureux, Charles H., University of Hawaii
Leeper, John R., Washington State University
Lloyd, Robert, University of Ohio
MacMillen, Richard E., University of California, Davis
Maka, Jean E., University of Hawaii
Matsunami, Lynn, University of Hawaii
McGurk, Linda L., University of Hawaii
Meeker, Joseph, University of Hawaii
Mi, Ming-Pi, University of Hawaii
Mitchell, Wallace C., University of Hawaii
Montgomery, Steve, University of Hawaii
Murakami, Gail, University of Hawaii
Myers, Barbara J., University of Hawaii
Nakahara, Larry M., University of Hawaii
Newman, Lester J., Portland State University, Oregon
Nishida, Toshiyuki, University of Hawaii
Paik, Yong K., Hanyang University, Korea
Parman, Terry T., National Park Service, Hawaii
Porter, John R., University of Hawaii
Radovsky, Frank J., Bernice P. Bishop Museum, Hawaii
Samuelson, G. Allan, Bernice P. Bishop Museum, Hawaii
Sattler, Klaus, British Museum of Natural History
Smith, Herman Eddie, University of Hawaii
Spatz, Günter O., Technische Universität München Institut fur
    Grünlandlehre
Sprenger, Daniel A., University of Hawaii
Steffan, Wallace A., Bernice P. Bishop Museum, Hawaii
Steiner, William W. M., University of Illinois

Stoner, Martin F., California State Polytechnic University
Sung, Kee Chang, Sung Kyun Kwan University, Korea
Tamashiro, Minoru, University of Hawaii
Tenorio, Joanne M., Bernice P. Bishop Museum, Hawaii
Tenorio, Joaquin, Bernice P. Bishop Museum, Hawaii
Tomich, P. Quentin, State Department of Health, Hawaii
Uchida, Grant, University of Hawaii
Van Riper, Charles, University of Hawaii
Wallace, Gordon D., Pacific Research Section, National Institute of
    Health, Honolulu
Warshauer, Frederick, University of Hawaii
Whittow, G. Causey, University of Hawaii
Wirawan, Nengah, University of Hawaii
Wirawan, Sarah, University of Hawaii
Yamashiro, Sandra, University of Hawaii
Yamashiro, Winifred Y., University of Hawaii

BIOLOGICAL PRODUCTIVITY IN UPWELLING ECOSYSTEMS

DIRECTORATE: Director, Richard C. Dugdale; Deputy Director, John J. Walsh

PRINCIPAL RESEARCH SITES:
  Peru Current
  Aegean Sea
  Canary Current
  Bay of Panama

DURATION OF PROGRAM: 1969-74

OBJECTIVES: To obtain sufficient quantitative understanding of the structure and dynamics of upwelling ecosystems to allow prediction of the consequences of perturbations of these and other marine ecosystems when these changes result primarily in enhanced circulation of nutrients into lighted regions of the sea. To apply the theoretical concepts so obtained to pilot aquacultural projects. To construct a simulation model of the Peruvian anchovy fishery, incorporating the dynamics of the nutrient concentration and primary productivity.

RESULTS: 50 papers in open literature (to June 1973).

PARTICIPANTS
Ballister, A., Instituto de Investigaciones Pesqueras, Barcelona, Spain
Barber, Richard T., Duke University. North Carolina
Bass, Perkins B., University of Washington
Blasco, D., Instituto de Investigaciones Pesqueras, Barcelona, Spain
Calvert, Stephen E., University of Edinburgh, Scotland
Cruzado, A., Instituto de Investigaciones Pesqueras, Barcelona, Spain
Davis, C., University of Washington
Frost, Bruce W., University of Washington
Goering, John J., University of Alaska
Gordon, Louis I., Oregon State University
Hopkins, Thomas S., Institute of Oceanographic and Fisheries Research
Kachel, N., University of Washington
Kelley, James C., University of Washington
MacIsaac, Jane J., University of Washington
Margalef, R., Instituto de Investigaciones Pesqueras, Barcelona, Spain
Murphy, Stanley R., University of Washington
O'Brien, James J., Florida State University
Packard, Theodore T., University of Washington
Paulik, Gerald J., (deceased), University of Washington
Pavlou, Spyros P., University of Washington
Pillsbury, R. Dale, Oregon State University

Piper, David Z., University of Washington
Pomeroy, Lawrence R., University of Georgia
Price, Norman B., University of Edinburgh, Scotland
Rowe, Gilbert T., Woods Hole Oceanographic Institution
Smith, Robert L., Oregon State University
Smyth, C. S., University of Washington
Szekielda, Karl H., University of Delaware
Thorne, Richard E., University of Washington
Walsh, John J., University of Washington
Whitledge, Terry E., University of Washington
Wiebe, William J., University of Georgia
Young, Dian L. K., Duke University, North Carolina

INTEGRATED PLANT PEST CONTROL

DIRECTORATE: Director, Carl B. Huffaker; Associate Director, Ray F. Smith

EXECUTIVE COMMITTEE:
Perry L. Adkisson
Stanley D. Beck
E. H. Glass
L. Dale Newsom
R. W. Stark

DURATION OF PROGRAM: 1970-

RESEARCH SITES: Numerous

OBJECTIVES: To study various tactics of pest control to develop the scientific basis for integrated methods of controlling populations of insect pests at non-economic densities in a manner that optimizes cost-benefit relations and minimizes environmental degradation.

RESULTS: 78 papers in open literature (to June 1973). Establishment of a cooperative multiuniversity, National Science Foundation, Environmental Protection Agency and U.S. Department of Agriculture program extending beyond the end of the IBP.

PARTICIPANTS
Adkisson, Perry L., Texas A&M University
Apple, J. L., North Carolina State University
Armbrust, Edward J., University of Illinois
Baker, D. M., ARS, USDA, Mississippi
Ball, Joe C., University of California, Berkeley
Beck, Stanley D., University of Wisconsin
Cartier, J. J., Canada Department of Agriculture
DeBach, Paul H., University of California, Riverside
Fisher, Theodore W., University of California, Riverside
Glass, E. H., New York State Agricultural Experiment Station
Guyer, Gordon E., Michigan State University
Headley, Joseph C., University of Missouri
Holling, C. S., University of British Columbia, Vancouver
Hoyt, Stanley C., Washington State University
Klassen, W., ARS, USDA, Washington, D.C.
Knipling, Edward F., ARS, USDA, Beltsville, Maryland
LeRoux, Edgar G., Canada Department of Agriculture
Martignoni, Mauro E., University of Oregon
Newsom, L. Dale, Louisiana State University
Rabb, Robert L., North Carolina State University

Riley, R. C., CSRS,USDA, Washington, D.C.
Shaw, W. C., ARS, USDA, Beltsville, Maryland
Stark, R. W., University of Idaho
Waters, William E., U.S. Forest Service, Washington, D.C.

Technical Panel
Craig, George B., Jr., University of Notre Dame
DeMichele, D. W., Texas A&M University
Ewing, Bland C., University of California, Berkeley
Headley, Joseph C., University of Missouri
Ignoffo, Carlo M., ARS, USDA, Columbia, Missouri
Lawson, Frank R., ARS, USDA (retired), Gainesville, Florida
Myers, Wayne L., Michigan State University
Norgaard, Richard B., University of California, Berkeley
Robbins, William E., ARS, USDA, Beltsville, Maryland
Rowe, Gordon A., University of California, Berkeley
Sailer, Reece I., University of Florida
Turnock, W. J., Canada Department of Agriculture

Alfalfa
Armbrust, Edward J., Illinois Natural History Survey
Allred, Keith R., Utah State University
Axtell, J. D., Purdue University, Indiana
Bartell, D., University of Illinois
Bohart, George E., Utah State University
Bouseman, John, University of Illinois
Brindley, W., Utah State University
Bremer, C., Cornell University, New York
Browning, D. R., Southern Illinois University
Bula, R. J., USDA, Lafayette, Indiana
Campbell, William F., Utah State University
Carmer, S. G., University of Illinois
Cherry, R., University of Illinois
Christensen, J., University of California, Berkeley
Cooley, N., Cornell University, New York
Cothran, Warren, University of California, Davis
Cowan, S., University of Illinois
Cruz, Jose, University of Illinois
Cubbin, C., University of California, Berkeley
David, Florence N., University of California, Riverside
Davis, Donald W., Utah State University
Day, William H., USDA, Moorestown, New Jersey
Dethier, Bernard E., Cornell University, New York
Dewey, Wade G., Utah State University

Diachun, Stephen, University of Kentucky
Dickason, E., University of Nebraska
Ditmars, S., Cornell University, New York
Etzel, L., University of California, Berkeley
Fick, Gary W., Cornell University, New York
Frosheiser, F., University of Minnesota
Giese, R., University of Illinois
Godfrey, George L., University of Illinois
Gonzalez, Daniel G., University of California, Riverside
Graffis, Don W., University of Illinois
Gutierrez, Andrew P., University of California, Berkeley
Gyrisco, George G., Cornell University, New York
Hanway, Donald G., University of Nebraska
Haws, Byron A., Utah State University
Helgesen, R. G., Cornell University, New York
Hildebrand, H., University of Illinois
Hittle, Carl N., University of Illinois
Hsiao, Ting H., Utah State University
Huber, Roger, Purdue University, Indiana
Jackobs, Joseph A., University of Illinois
Jensen, R. Z., Utah State University
Jordan, W., University of California, Berkeley
Kehr, William R., University of Nebraska
Kindler, S. D., University of Nebraska
Klostermeyer, L., University of Nebraska
Koehler, Carlton S., University of California, Berkeley
Kogan, Marcos, Illinois Natural History Survey
Kotek, R., University of Illinois
Kugler, J., University of Nebraska
Latheef, M. A., University of Kentucky
Lee, K., University of California, Berkeley
Lehman, William F., University of California, Davis
Lichtenstein, E. Paul, University of Illinois
Lin, B., Cornell University, New York
Lowe, Carl C., Cornell University, New York
Maddox, J. V., Illinois Natural History Survey
Manglitz, George R., University of Nebraska
McAllister, De Vere R., Utah State University
Miller, D. A., University of Illinois
Mitchell, George E., Jr., University of Kentucky
Montgomery, V., University of Kentucky
Moore, G., University of California, Berkeley
Morrison, W., University of Kentucky
Murphy, R. C., Cornell University, New York

Nebeker, T., Utah State University
Nichols, M., University of Illinois
Norgaard, Richard B., University of California, Berkeley
Nye, William P., Utah State University
Ogden, R. L., University of Nebraska
Palmer, L., University of Nebraska
Pass, B. C., University of Kentucky
Pederson, Marion W., Utah State University
Perkins, William, University of Illinois
Peterson, B., University of Illinois
Pienkowski, Robert L., Virginia Polytechnic Institute
Price, Peter W., University of Illinois
Ratcliffe, R., ARS, USDA, Beltsville, Maryland
Regev, U., University of California, Berkeley
Roberts, Stephen J., University of Illinois
Rose, M., University of Illinois
Ruesink, W. G., University of Illinois
Sant, M., Utah State University
Sevacherian, Vahram, University of California, Riverside
Shoemaker, Christine, Cornell University, New York
Smith, G., Cornell University, New York
Stanford, E. H., University of California, Davis
Stannard, Lewis J., Illinois Natural History Survey
Stern, Vernon M., University of California, Riverside
Stoltz, R., University of California, Berkeley
Summers, Charles G., University of California, Berkeley
Swanson, E. R., University of Illinois
Tauber, M., Cornell University
Taylor, C., University of Illinois
Taylor, Norman L., University of Kentucky
Taylor, R., University of Illinois
Taylor, Timothy H., University of Kentucky
Templeton, William C., Jr., University of Kentucky
Thompson, Lafayette, Jr., University of Kentucky
Thurston, Richard B., University of Kentucky
Torchio, P. F., USDA, Logan, Utah
Tucker, Roy E., University of Kentucky
van den Bosch, Robert A., University of California, Berkeley
Vignarajah, I., Cornell University, New York
White, C. E., University of Illinois
Wiggins, W., University of Kentucky
Wilson, Mark C., Purdue University, Indiana
Wolf, D. D., Virginia Polytechnic Institute
Wright, Madison J., Cornell University, New York

Young, A., University of Kentucky
Youssef, N., Utah State University

Citrus
Allen, J., University of Florida
Avis, Don, Sunkist Growers, Inc., Van Nuys, California
Ball, Joe C., University of California, Berkeley
Baranowski, Richard M., University of Florida
Black, J. Hodge, University of California Extension, Bakersfield
Brewer, R., University of California, Riverside
Brooks, Robert F., University of Florida
Burns, R., University of California Extension, Ventura
Bush, J., Texas A&M University
Calavan, Edmond, University of California, Riverside
Chambers, D. L., USDA, Gainesville, Florida
Cintron, R., Texas A&M University
Collins, F., University of Florida
David, Florence N., University of California, Riverside
Deal, Andrew S., University of California, Riverside
Dean, Herbert A., Texas A&M University
Dietrick, E. J., Rincon-Vitova Insectaries, Inc., Rialto, California
Doutt, Richard L., University of California, Parlier
Dulmage, Howard T., USDA, Brownsville, Texas
Duncan, W., Superior Farming, Bakersfield, California
Elder, C., University of California, Riverside
Elmer, H., University of California, Riverside
Ewart, W. H., University of California, Riverside
Ewing, Bland C., University of California, Berkeley
Fisher, Theodore W., University of California, Riverside
Flaherty, Donald L., University of California Extension, Visalia
Francis, H., University of California, Riverside
Gorden, J., Pest Management Consultant, Exeter, California
Green, Karen, University of California, Davis
Hagen, Kenneth S., University of California, Berkeley
Hart, W. G., USDA, Weslaco, Texas
Hartley, H., Texas A&M University
Headley, Joseph C., University of Missouri
Hench, Kenneth W., University of California Extension, Bakersfield
Huftile, M., University of California, Riverside
Jeppson, Lee R., University of California, Riverside
Jones, William L., Texas A&M University
Jordan, L. S., University of California, Riverside
Kennett, Charles E., University of California, Berkeley
Luck, Robert F., University of California, Riverside

Maxwell, N., Texas A&M University
McCoy, Clayton W., University of Florida
McLain, R., Visalia, California
McMurtry, James A., University of California, Riverside
Menke, W. W., University of Florida
Moreno, D., USDA, Riverside, California
Mostafa, Abdel, University of California, Riverside
Opitz, Karl, University of California Extension, Parlier
Orth, R., University of California, Riverside
Oster, George F., University of California, Berkeley
Pehrson, J. E., University of California Extension, Visalia
Pennington, N., Fillmore Citrus Protection District
Platt, R., University of California, Riverside
Quezada, Jose, University of California Extension, Visalia
Reed, David H., USDA, Riverside, California
Riddle, Don, Gilkey Farms, Orosi, California
Riehl, Louis A., University of California, Riverside
Seils, A., Texas A&M University
Selhime, Allen G., USDA, Orlando, Florida
Sevacherian, Vahram, University of California, Riverside
Shaw, John G., USDA, Riverside, California
Shiells, J., Shiells Ranch Company
Snyder, J. Herbert, University of California, Davis
Surber, Delbert E., Superior Farming, Bakersfield, California
Villalon, Ben, Texas A&M University
Warner, S., University of California, Riverside
Waters, R., Perma Rain Corporation, Lindsey, California
Whitcomb, Willard H., University of Florida
White, W., University of California, Riverside
Whitlock, M., Woodcrest Citrus Project, Riverside, California
Wiley, W., University of California, Berkeley

Cotton
Adkisson, Perry L., Texas A&M University
Akesson, Norman B., University of California, Davis
Allen, Robert T., University of Arkansas
Andrews, G. L., Mississippi State University
Baker, D. N., ARS, USDA, State College, Mississippi
Billingsley, R. V., Texas A&M University
Bird, Luther S., Texas A&M University
Bottrell, Dale G., Texas A&M University
Brook, Theodore S., Mississippi State University
Brown, L., Mississippi State University
Bull, D. L., ARS, USDA, College Station, Texas

Burke, H. R., Texas A&M University
Casey, B., Texas A&M University
Chambers, Howard W., Mississippi State University
Colwick, Rex F., USDA, State College, Mississippi
Coppedge, J. R., ARS, USDA, College Station, Texas
Cowan, C. B., ARS, USDA, Waco, Texas
Curry, H. L., Texas A&M University
David, Florence N., University of California, Riverside
DeMichele, D. W., Texas A&M University
Denton, W. H., University of California, Riverside
Douglas, Alvin G., Mississippi State University
Falcon, Louis A., University of California, Berkeley
Frazier, James L., Mississippi State University
Gannaway, J., Texas A&M University
Gonzalez, Daniel G., University of California, Riverside
Graham, H. M., USDA, Brownsville, Texas
Grimes, Donald W., University of California Extension, Parlier
Gutierrez, Andrew P., University of California, Berkeley
Hagen, Kenneth S., University of California, Berkeley
Hamill, J. G., USDA, State College, Mississippi
Hanna, Ralph L., Texas A&M University
Hardee, D. D., USDA, State College, Mississippi
Harding, James A., Texas A&M University
Harris, F. Aubrey, Mississippi State University
Holder, David G., Mississippi State University
Hueth, D., University of California, Berkeley
Hurt, V. G., Mississippi State University
Hyer, Angus, USDA, Shafter, California
Jenkins, Johnie N., USDA, State College, Mississippi
Kay, R. D., Texas A&M University
Lacewell, Ronald D., Texas A&M University
Laster, Marion L., Mississippi State University
Lee, K., University of California, Berkeley
Leigh, Thomas F., University of California, Davis
Lincoln, Charles G., University of Arkansas
Lipps, Karen L., University of California, Riverside
Lloyd, Edwin P., Mississippi State University
Lukefahr, Maurice J., USDA, Brownsville, Texas
Martin, Dial F., USDA, Stoneville, Mississippi
Maxwell, Fowden G., Mississippi State University
McLaughlin, Roy E., USDA, State College, Mississippi
Megehee, J. H., Mississippi State University
Moody, D., Texas A&M University
Nemec, Stanley J., Texas A&M University
Niles, George A., Texas A&M University

Norgaard, Richard B., University of California, Berkeley
Parker, M., Mississippi State University
Parrott, William L., USDA, State College, Mississippi
Parvin, D., Mississippi State University
Phillips, Jacob R., University of Arkansas
Plapp, Frederick W., Jr., Texas A&M University
Powell, Robert D., Texas A&M University
Regev, U., University of California, Berkeley
Reynolds, Harold T., University of California, Riverside
Ridgeway, Richard L., USDA, College Station, Texas
Sabbe, W., University of Arkansas
Schuster, M. F., Texas A&M University
Sevacherian, Vahram, University of California, Riverside
Sharpe, P., Texas A&M University
Shaunak, Krishan K., Mississippi State University
Sikorowski, Peter P., Mississippi State University
Skeith, R. W., University of Arkansas
Slosser, J. E., University of Arkansas
Sterling, Winfield L., Texas A&M University
Stern, Vernon M., University of California, Riverside
Talpaz, H., Texas A&M University
Threadgill, Ernest D., Mississippi State University
Tugwell, Noel P., Jr., University of Arkansas
van den Bosch, Robert A., University of California, Berkeley
Waddle, Bradford A., University of Arkansas
Walker, James K., Jr., Texas A&M University
Wilkes, Lambert H., Texas A&M University
Wiley, W., University of California, Berkeley
Wilson, Clifton A., Mississippi State University
Wolfenbarger, D. D., ARS, USDA, Brownsville, Texas
Yearian, William C., Jr., University of Arkansas

Pine (Bark Beetles)
Adams, David L., University of Idaho
Amman, Gene E., U.S. Forest Service, Ogden, Utah
Anderson, Lee, University of Idaho
Anderson, Roger F., Duke University, North Carolina
Baldwin, Paul H., Colorado State University
Barras, S. J., U.S. Forest Service, Pineville, Louisiana
Bedard, William D., U.S. Forest Service, Berkeley, California
Bennett, William H., U.S. Forest Service, Pineville, Louisiana
Berryman, Alan A., Washington State University
Bruce, Richard W., Washington State University
Burnell, D. G., Washington State University

Chapman, Roger A., Duke University, North Carolina
Cobb, Fields W., Jr., University of California, Berkeley
Cole, Dennis M., U.S. Forest Service, Bozeman, Montana
Cole, Walter E., U.S. Forest Service, Ogden, Utah
Coster, J., Texas A&M University
Coulson, Robert N., Texas A&M University
Dahlsten, Donald L., University of California, Berkeley
Dalleske, R. L., U.S. Forest Service, Berkeley, California
Deitschman, Glenn H., U.S. Forest Service, Moscow, Idaho
DeMars, Clarence J., U.S. Forest Service, Berkeley, California
Dingle, Richard W., Washington State University
Ditweiler, C. D., Washington State University
Drooz, Arnold T., U.S. Forest Service, Research Triangle Park, North Carolina
Echols, H. W., Mississippi Forest Commission, Jackson
Ewing, Bland C., University of California, Berkeley
Furniss, Malcolm M., University of California, Berkeley
Galbraith, J. M., U.S. Forest Service, Ogden, Utah
Gilkerson, Raymond A., Washington State University
Godfrey, Erik B., University of Idaho
Gustafson, R., U.S. Forest Service, San Francisco, California
Hamilton, David A., Jr., U.S. Forest Service, Moscow, Idaho
Heikkenen, Herman J., Virginia Polytechnic Institute
Heller, Robert, U.S. Forest Service, Berkeley, California
Helms, John A., University of California, Berkeley
Klein, W. H., U.S. Forest Service, Ogden, Utah
Knauer, Ken, U.S. Forest Service, Atlanta, Georgia
Knover, K. H., U.S. Forest Service, Atlanta, Georgia
Landgraf, A. E., U.S. Forest Service, Atlanta, Georgia
Leaphart, Charles D., U.S. Forest Service, Moscow, Idaho
Leuschner, William A., Virginia Polytechnic Institute
Lorio, Peter L., Jr., U.S. Forest Service, Pineville, Louisiana
Lotan, James E., U.S. Forest Service, Bozeman, Montana
Lyon, C. J., U.S. Forest Service, Missoula, Montana
McCambridge, William F., U.S. Forest Service, Fort Collins, Colorado
Michalson, Edgar L., University of Idaho
Moser, John C., U.S. Forest Service, Pineville, Louisiana
Myers, Clifford A., Jr., U.S. Forest Service, Fort Collins, Colorado
Newton, Carlton M., Virginia Polytechnic Institute
Olson, Donald E., University of Idaho
O'Regan, William G., U.S. Forest Service, Berkeley, California
Orr, Peter N., U.S. Forest Service, Portland, Oregon
Parmeter, J. R., Jr., University of California, Berkeley
Payne, Thomas L., Texas A&M University

Pfister, Robert D., U.S. Forest Service, Missoula, Montana
Pienaar, Leon V., Washington State University
Pierce, Donald A., U.S. Forest Service, Pineville, Louisiana
Pierce, D. M., U.S. Forest Service, Upper Darby, Pennsylvania
Pitman, G., Boyce Thompson Institute, California
Rauch, Peter A., University of California, Berkeley
Rauchensberger, John L., U.S. Forest Service, Asheville, North Carolina
Reynolds, Marion R., Jr., Virginia Polytechnic Institute
Rigas, Anthony L., University of Idaho
Rigas, Harriet B., Washington State University
Roe, Arthur L., U.S. Forest Service (retired), Ogden, Utah
Sartwell, Charles, Jr., U.S. Forest Service, Corvallis, Oregon
Schenk, John A., University of Idaho
Schmidt, Wyman C., U.S. Forest Service, Missoula, Montana
Schmitz, Richard F., U.S. Forest Service, Moscow, Idaho
Shearer, Raymond C., U.S. Forest Service, Missoula, Montana
Shew, Richard L., Washington State University
Stage, Albert T., U.S. Forest Service, Moscow, Idaho
Stark, R. W., Director, University of Idaho
Stevens, Robert E., U.S. Forest Service, Fort Collins, Colorado
Sullivan, Alfred D., Virginia Polytechnic Institute
Thatcher, Robert C., U.S. Forest Service, Pineville, Louisiana
Thomas, J., University of Idaho
Waters, William E., U.S. Forest Service, Washington, D.C.
Wood, David L., University of California, Berkeley
Yandle, D. O., Duke University, North Carolina

Pome and Stone Fruits
Akesson, Norman B., University of California, Davis
Anthon, Edward W., Washington State University
Asquith, Dean, Pennsylvania State University
Ball, Joe C., University of California, Berkeley
Batiste, William C., University of California, Berkeley
Baugher, D., Pennsylvania State University
Benson, Nels R., Washington State University
Bode, William M., Pennsylvania State University
Brown, Dillon S., University of California, Davis
Brunner, J., Washington State University
Burts, Everett C., Washington State University
Butt, William A., ARS, USDA, Bethesda, Maryland
Caltagirone, Leopoldo E., University of California, Berkeley
Colburn, R., Pennsylvania State University
Covey, R., Washington State University
Croft, Brian A., Michigan State University

Daum, Donald R., Pennsylvania State University
Davidson, A., University of California, Berkeley
Davis, Clarence S., University of California, Berkeley
Dibble, John E., University of California, Berkeley
Dress, P. E., Pennsylvania State University
Eves, J., Washington State University
Ewing, Bland C., University of California, Berkeley
Falcon, Louis A., University of California, Berkeley
Greene, George M., II, Pennsylvania State University
Gutierrez, Andrew P., University of California, Berkeley
Hagen, Kenneth S., University of California, Berkeley
Harsh, Stephen B., Michigan State University
Harwood, Robert F., Washington State University
Haynes, Dean L., Michigan State University
Howitt, Angus J., Michigan State University
Hoying, S., Michigan State University
Hoyt, Stanley C., Director, Washington State University
Hull, Jerome, Jr., Michigan State University
Jones, Alan L., Michigan State University
Kelley, Bernard Wayne, Pennsylvania State University
Ketchie, Del O., Washington State University
Koenig, Herman E., Michigan State University
Lewis, Fred H., Pennsylvania State University
Logan, J., Washington State University
Meyer, R., Michigan State University
Moffitt, Harold R., USDA, Yakima, Washington
Mowery, P., Pennsylvania State University
Nakashima, M., Michigan State University
Ogawa, Joseph M., University of California, Davis
Okumura, George, University of California, Berkeley
Olsen, Kenneth R., ARS, USDA, Wenatchee, Washington
Oster, George F., University of California, Berkeley
Putnam, A., Michigan State University
Rice, Richard E., University of California, Davis
Riedl, H. W., Michigan State University
Rodriguez, J., Michigan State University
Roelofs, W., University of California, Berkeley
Rowe, Gordon A., University of California, Berkeley
Russell, T., Washington State University
Smith, Samuel H., Pennsylvania State University
Stahly, Edward A., ARS, USDA, Wenatchee, Washington
Stevens, C., USDA, Mount Pleasant, Michigan
Strombler, V., Del Monte Corporation
Tanigoshi, Lynell, Washington State University

Tette, J. P., Zoecon Corporation, Palo Alto, California
Thomas, J. D., Pennsylvania State University
Thompson, W. W., Michigan State University
Tummala, Lal, Michigan State University
Vanderbrink, Ceel, Michigan State University

Soybean

Allen, George E., Florida Technological University
Armbrust, Edward J., Illinois Natural History Survey
Bernard, Richard L., ARS, USDA, Urbana !llinois
Birchfield, Wray B., Louisiana State University
Bouseman, John, University of Illinois
Bradley, J. R., Jr., North Carolina State University
Brim, Charles A., North Carolina State University
Brooks, Wayne M., North Carolina State University
Callahan, P., University of Florida
Campbell, William V., North Carolina State University
Carlson, George A., North Carolina State University
Carner, Gerald R., Clemson University, South Carolina
Farthing, Barton R., Louisiana State University
Ford, B., University of Illinois
Godfrey, George L., University of Illinois
Graves, Jerry B., Louisiana State University
Greene, Gerald L., University of Florida
Guthrie, Frank E., North Carolina State University
Hammond, Abner M., Jr., Louisiana State University
Hartwig, Edgar E., ARS, USDA, Stoneville, Mississippi
Hebert, Teddy T., North Carolina State University
Herzog, D., Louisiana State University
Hodgson, Ernest, North Carolina State University
Horn, Norman L., Louisiana State University
Jensen, R. L., Louisiana State University
Kish, L., University of Florida
Kogan, Marcos, Illinois Natural History Survey
Larson, A. L., Louisiana State University
Leppla, N., University of Florida
Maddox, J. V., Illinois Natural History Survey
Martin, Dial F., ARS, USDA, Stoneville, Mississippi
Maxwell, James D., Clemson University, South Carolina
Mays, D., University of Florida
Menke, W. W., University of Florida
Miller, Marth P., University of Illinois
Miller, P., North Carolina State University
Miner, Floyd D., University of Arkansas

Mueller, A. J., University of Arkansas
Neunzig, Herbert H., North Carolina State University
Newsom, L. Dale, Director, Louisiana State University
Price, Peter W., University of Illinois
Rabb, Robert L., North Carolina State University
Rawlings, J. O., North Carolina State University
Roberts, Stephen J., University of Illinois
Rogers, N. L., Louisiana State University
Ross, John P., North Carolina State University
Rudd, W. G., Louisiana State University
Ruesink, W. G., University of Illinois
Schillinger, J., Clemson University, South Carolina
Scott, Howard A., University of Arkansas
Sheets, Thomas J., North Carolina State University
Shepard, B. M., Clemson University, South Carolina
Sprenkel, Richard K., North Carolina State University
Srinivasan, Vahram R., Louisiana State University
Stannard, Lewis J., Illinois Natural History Survey
Sullivan M. J., Clemson University, South Carolina
Tanigoshi, Lynell, Washington State University
Tugwell, Noel P., Jr., University of Arkansas
Turnipseed, Samuel C., Clemson University, South Carolina
Van Duyn, J., North Carolina State University
Waldbauer, Gilbert P., University of Illinois
Walters, Herbert J., University of Arkansas
Whitcomb, Willard H., University of Florida
Wiegmann, F. H., Louisiana State University
Williams, C., Louisiana State University
Yearian, William C., Jr., University of Arkansas
Young, Seth Y., III, University of Arkansas

BIOLOGY OF HUMAN POPULATIONS AT HIGH ALTITUDES

DIRECTORATE: Director, Paul T. Baker, Pennsylvania State University

RESEARCH SITES: Nuñoa and several other cities and towns in the southern Peruvian Andes; Sherpa and Tibetan populations in Nepal

DURATION OF PROGRAM: 1968-75 (but expansion on pre-1968 work)

OBJECTIVES: To obtain an understanding of how human populations at high altitude adapt to their environment and of what happens to their biology when they migrate downward.

RESULTS: 80 papers in open literature (to June 1973), one book, "Man in the Andes: The High Altitude Quechua," in US/IBP Synthesis Series. Major contributions to the international synthesis volume "The Biology of High Altitude Peoples."

PARTICIPANTS
Demography
Abelson, Andrew, Pennsylvania State University
DeJong, Gordon F., Pennsylvania State University
Dutt, James S., Pennsylvania State University
Hoff, Charles J., University of Oregon
Spector, Richard, Pennsylvania State University

Growth and Child Development
Baker, Thelma, Pennsylvania State University
Frisancho, A. Roberto, University of Michigan
Haas, Jere, University of Massachusetts
Hoff, Charles J., University of Oregon
McGarvey, Steven T., Pennsylvania State University
Pawson, I. Guy, G. W. Hooper Foundation, University of California, San Francisco

Cardiovascular Physiology
Watt, Edward, Sports Medicine Clinic, Atlanta, Georgia

Environmental Physiology
Brewer, George J., University of Michigan
Buskirk, Elsworth R., Pennsylvania State University
Eaton, John, University of Michigan
Hanna, Joel M., University of Hawaii
Howley, E. T., University of Tennessee
Kollias, James, Pennsylvania State University

Larsen, Robert M., University of Wisconsin
Little, Michael A., State University of New York at Binghamton

Respiratory Physiology
Austin, Donald M., Southern Methodist University
Lahiri, Sukhamay, University of Pennsylvania
Velasquez, Tulio, University of San Marcos, Peru

Work Capacity
Mazess, Richard B., University of Wisconsin
Weitz, Charles A., Temple University, Pennsylvania

Nutrition
Gursky, M. J., Pennsylvania State University
Mendez, Jose, Pennsylvania State University
Picon-Reategui, Emilio, University of San Marcos, Peru

General Ecology
Thomas, R. Brooke, Cornell University, New York
Winterhalder, Bruce, Cornell University, New York

Infectious Disease, Hematology, and Health
Beall, Cynthia, Pennsylvania State University
Garruto, Ralph M., National Institute of Neurological Disease and Stroke
Way, Anthony, Texas Tech University School of Medicine

Social Structure
Escobar, Gabriel, Pennsylvania State University

POPULATION GENETICS OF SOUTH AMERICAN INDIANS

DIRECTORATE: Director, James V. Neel

PRINCIPAL RESEARCH SITES: Yanomama villages along Venezuelan-Brazilian border

DURATION OF PROGRAM: 1968-74 (but continuation of pre-1968 work)

OBJECTIVES: To learn the tempo of human evolution as measured by the degree of divergence that has arisen between Indian groups since their arrival in the Americas. For those Indian groups which still exist in a pre-Columbian state, to learn what significant biological parameters have affected their evolution. To learn what patterns of social and biological adaptation emerge as primitive Indian groups are subjected to cultural change. To learn the extent to which cultural and social practices (marriage, settlement pattern, intergroup hostility) affect the biological and demographic characteristics of the populations.

RESULTS: 79 papers in the open literature (to June 1973)

PARTICIPANTS
Laboratory Studies
Arends, T., Instituto Venezolano de Investigaciones Cientificas, Venezuela
Ayres, Manuel, Universidade Federal do Para, Belem, Brazil
Bergold, G. H., Instituto Venezolano de Investigaciones Cientificas, Venezuela
Bloom, Arthur D., University of Michigan
Casey, Helen, U.S. Department of Health, Education, and Welfare
Eveland, Warren C., University of Michigan
Gershowitz, Henry, University of Michigan
Layrisse, Miguel, Instituto Venezolano de Investigaciones Cientificas, Venezuela
LeQuesne, Philip, University of Michigan
Shreffler, Donald C., University of Michigan
Tanis, Robert J., University of Michigan
Tashian, Richard E., University of Michigan
Weitkamp, Lowell R., University of Rochester

Ethnology
Asch, Timothy, Cambridge, Massachusetts
Chagnon, Napoleon, Pennsylvania State University
Migliazza, Ernest, University of Maryland

Medical Studies
Oliver, William E., University of Michigan

Computer Simulation
Griffith, Ronald, University of Michigan
Li, Francis, University of Michigan

Mathematical Genetics
Neel, James V., University of Michigan
Rothman, Edward D., University of Michigan
Salzano, Francisco M., Universidade de Rio Grande do Sul, Brazil
Schull, William J., University of Texas, Houston
Spielman, Richard S., University of Michigan
Ward, Richard H., University of Michigan
Weiss, Kenneth M., University of Michigan

Other Participants
Ferrell, Robert E., University of Michigan
Jensen, Louis, University of Michigan

INTERNATIONAL STUDY OF CIRCUMPOLAR PEOPLES

International Study of Eskimos Subprogram

DIRECTORATE: Director, Frederick A. Milan

RESEARCH SITES:
  Principal Site
    Wainwright, Alaska
  Other Sites
    Point Hope
    Barrow
    Kaktovik (Barter Island)
    Anaktuvuk Pass

DURATION OF PROGRAM: 1968-71

OBJECTIVES: To describe biological and behavioral processes responsible for the successful adaptation and perpetuation of breeding isolates of Eskimos in northern Alaska.

RESULTS: 239 papers in open literature (to June 1976), one volume, "The Eskimo of Northwestern Alaska," in US/IBP Synthesis Series (in preparation).

PARTICIPANTS
Anthropometry, Anthroposcopy, and Photography
Hudson, H., Indiana University
Jamison, Paul L., Indiana University
Laughlin, William S., University of Connecticut
Meier, Robert J., Indiana University
Pawson, I. G., University of California, Los Angeles
Pawson, S., University of California, Los Angeles
Zegura, S. L., University of Arizona

Behavior Studies
Bock, R. Darrell, University of Chicago
Brøsted, J., University of Copenhagen, Denmark
Feldman, Carol S., University of Chicago
Fitzgerald, W., University of Michigan Dental School
Foulks, Edward F., University of Pennsylvania
Kolakowski, Donald, University of Connecticut
McLean, D., University of Chicago
Sternbach, Ruth, University of Connecticut

Chronobiology
Bohlen, B., Harvard Law School, Massachusetts
Bohlen, J. G., University of Minnesota Medical School
Bohlen, S., Minneapolis, Minnesota
Halberg, Franz, University of Minnesota Medical School

Dental Examinations, X-Rays, and Casts
Dahlberg, A. A., Zoller Memorial Dentral Clinic, University of Chicago
Dahlberg, T., University of Chicago
Forti, T., University of Chicago
Kulesz, M., University of Chicago
Likens, D., University of Chicago
Mayhall, J. T., University of Toronto, Canada
Merbs, C. F., Arizona State University
Owen, D. G., University of Maryland
Walker, P., University of Chicago

Exercise Physiology
Di Prampero, P., Milan, Italy
Fritts, R. W., University of Wisconsin
Hogan, P., State University of New York at Buffalo
Rennie, Donald W., State University of New York at Buffalo
Sinclair, Lynn, State University of New York at Buffalo
Washburn, R., State University of New York at Buffalo
Wilson, D., State University of New York at Buffalo
Wilson, M., State University of New York at Buffalo

Epidemiology, Clinical Chemistry, and Population Genetics
Allen, Fred, Jr., New York Blood Center
Boettcher, B., University of Flinders, Western Australia
Bosman, D., University of Wisconsin
Laessig, R. H., University of Wisconsin
Moore, Mary Jane, University of California, San Diego
Osborne, Richard H., University of Wisconsin
Palczer, R., University of Alaska
Pauls, Frank P., Alaska Division of Public Health, Juneau
Thompson, W., University of Wisconsin

Genealogy and Population History
MacLean, E. Ahgiak, University of Alaska
Milan, Frederick A., University of Alaska

240   W. F. Blair

Medical Examinations, ECG's, and X-Rays
Bates, T., University of Oregon
Dotter, Charles T., University of Oregon Medical School
Forsius, Henrik, University Eye Hospital, Oulu, Finland
Griswold, H., University of Oregon Medical School
Haraldson, S., University of Gothenburg, Sweden
Lewin, T., University of Gothenburg, Sweden
Mather, W., University of Wisconsin
Mazess, Richard B., University of Wisconsin
Pegg, Jack E., University of Oregon
Rice, R. B., Fredericksburg, Virginia
Robinhold, D., University of Colorado
Way, Anthony B., Texas Tech University School of Medicine
Wishart, D., University of Oregon Medical School

Nutrition
Bell, R. R., University of Illinois
Bergan, J. G., University of Illinois
Colbert, M. J., University of Illinois
Draper, Harold H., University of Illinois
Feldman, S. A., Northwestern University, Illinois
Goad, W., U.S. Army Medical and Nutritional Research Laboratory
Heller, Christine A., U.S. Public Health Service (retired)
Hursh, Laurence M., University of Illinois
Mann, George V., Vanderbilt University, Tennessee
Sauberlich, Howard E., U.S. Army Medical and Nutritional Research Laboratory
Wo, Catherine W., University of Illinois

International Study of Aleuts Subprogram

DIRECTORATE: Director, William S. Laughlin

PRINCIPAL RESEARCH SITES:
A. Archaeological
    Nikolski Bay region, Umnak Island, eastern Aleutian Islands
    Anangula Island
    Chaluka
    Ogalodagh
    Lake Baikal, U.S.S.R.
    Novosibirsk, U.S.S.R.
B. Human Biology
    Atka
    Akutan
    Nikolski
    Unalaska

DURATION OF PROGRAM: 1970-75

OBJECTIVES: To investigate present and past human populations of the Aleutian Islands ecosystem within an evolutionary framework.

PRINCIPAL DATA ANALYSIS CENTER: Laboratory of Biological Anthropology, Department of Biobehavioral Sciences, The University of Connecticut, Storrs, Connecticut

RESULTS: 14 papers in open literature (to July 1973), one book, "Aleut Population and Ecosystem Studies," in US/IBP Synthesis Series (in preparation), and several subsequent articles, some in Russian in Soviet scientific journals and books.

PARTICIPANTS: This investigation of the Aleut ecosystem has been a multidisciplinary project in human adaptability. Participants came from Denmark, France, and the Soviet Union, as well as the United States, to study problems related to their particular areas of expertise. Many of them worked in the Aleutians during more than one year and made research contributions which fall into more than one of the categories listed below. Participants have been listed in only one category although they made multiple contributions to the project. (The list includes two scientists and four secretaries who did not go to the field.)

Archaeology
Aigner, Jean S., University of Connecticut
Atkinson, Alice, Hunter College, New York
Bieber, Alan, University of Connecticut

Chatters, Barbara, University of Connecticut
Chatters, James, University of Connecticut
Elbaum, Kathy, University of Chicago
Finney, Bruce, Seattle, Washington
Frechette, Gregory, University of Connecticut
Frøhlich, Bruno, University of Connecticut
Fullem, Bruce, University of Connecticut
Ginsburg, Faye, University of Connecticut
Greene, John D., University of Connecticut
Jørgensen, B., University of Copenhagen, Denmark
Konigsberg, Diane, University of Connecticut
Kopjanski, David, University of Connecticut
Laughlin, Ruth, University of Connecticut
Laughlin, Sara, University of Connecticut
Lippold, Lois, San Diego State University
MacDowell, Mary, University of Connecticut
Myers, Dan, Eastern Washington State College
Rohrbach, David, University of Connecticut
Veltre, Douglas, University of Connecticut
Veltre, Mary, University of Connecticut
Wyss, James, University of Kentucky
Yesner, David, University of Connecticut

Ecosystem Analysis
Dubos, Robert, University of Connecticut
Gray, Norman, University of Connecticut
Hett, Joan M., University of Washington, Seattle
Love, Gordon, University of Connecticut

Ethnography-Demography
Holmes, Beverly, University of Connecticut
Robert-Lamblin, Joëlle, Centre des Recherches Anthropologiques, Paris
Sternbach, Ruth, University of Connecticut

Geology
Abraham, Paul, University of Connecticut
Black, Robert F., University of Connecticut
Funk, James, University of Connecticut
Galicki, Alan, Wittenberg University, Ohio
Meyers, Thomas, University of Connecticut
Thompson, Bruce, Princeton University, New Jersey

Human Biology
Alexander, Fred, University of Pennsylvania Medical Center
Beman, Susan, University of Connecticut
Giacobini, Ezio, University of Connecticut
Milman, Evelyn, University of Connecticut
Poole, Andrew, University of Connecticut
Poole, Deirdre, University of Connecticut
Wolf, Susan, University of Connecticut

U.S.-U.S.S.R. Joint Research
   Soviet team on Anangula Island, 1974
Derevyanko, A. P., Institute of History, Philology, and Philosophy, Siberian Division of the Academy of Sciences, Novosibirsk, U.S.S.R.
Konopatski, A., Institute of History, Philology, and Philosophy, Siberian Division of the Academy of Sciences, Novosibirsk, U.S.S.R.
Larichev, V., Institute of History, Philology, and Philosophy, Siberian Division of the Academy of Sciences, Novosibirsk, U.S.S.R.
Okladnikov, A. P., Institute of History, Philology, and Philosophy, Siberian Division of the Academy of Sciences, Novosibirsk, U.S.S.R.
Vasilievsky, R., Institute of History, Philology, and Philosophy, Siberian Division of the Academy of Sciences, Novosibirsk, U.S.S.R.
   Siberian research team, 1975: Lake Baikal
Atseev, I. V., Institute of History, Philology, and Philosophy, Siberian Division of the Academy of Sciences, Novosibirsk, U.S.S.R.
Campbell, J., University of New Mexico
Clark, D., National Museums of Canada
Derevyanko, A. P., Institute of History, Philology, and Philosophy, Siberian Division of the Academy of Sciences, Novosibirsk, U.S.S.R.
Harper, A. B., University of Connecticut
Hopkins, D., U.S. Geological Survey
Konopatski, A., Institute of History, Philology, and Philosophy, Siberian Division of the Academy of Sciences, Novosibirsk, U.S.S.R.
Laughlin, W. S., University of Connecticut
Okladnikov, A. P., Institute of History, Philology, and Philosophy, Siberian Division of the Academy of Sciences, Novosibirsk, U.S.S.R.
Okladnikova, E., Institute of History, Philology, and Philosophy, Siberian Division of the Academy of Sciences, Novosibirsk, U.S.S.R.
Troitsky, S., Institute of Geology and Geophysics, Siberian Division of the Academy of Sciences, Novosibirsk, U.S.S.R.

Other Participants
Gordon, Selma, University of Connecticut
Jones, Robert, Alaskan Wildlife Refuge

Kelley, John, University of Alaska
Kelley, Kathleen, University of Connecticut
Kolakowski, Donald, University of Connecticut
Krukoff, Oxenia, Nikolski, Alaska
Placek, Catharine L., University of Connecticut
Rosenthal, Fay, University of Connecticut
Stuckenrath, Robert, Radiocarbon Laboratory, Smithsonian Institution

Appendix B

Coordinated Research Programs of the US/IBP

AEROBIOLOGY

DIRECTORATE: Director, William S. Benninghoff, 1969-72; Program Coordinator, Robert L. Edmonds, 1971-72; Director, Robert L. Edmonds, 1972-75

DURATION OF PROGRAM: 1969-75

OBJECTIVES: To promote and coordinate nationally and internationally oriented research projects in critical areas of aerobiology.

RESULTS: 29 papers in open literature (to June 1973); one book, "Aerobiology," for US/IBP Synthesis Series (in preparation); formation of an "International Association for Aerobiology" in 1974.

PARTICIPANTS
Akers, Thomas G., Naval Biomedical Research Lab, Oakland, California
Bell, Nancie L., U.S. Coast Guard, San Francisco, California
Banaszak, E. F., St. Luke's Hospital, Milwaukee, Wisconsin
Bedard, William D., U.S. Forest Service, Berkeley, California
Benninghoff, William S., University of Michigan
Bernstein, I. Leonard, University of Cincinnati, Ohio
Brown, R. Malcolm, Jr., University of North Carolina
Burleigh, James R., Chico State University, California
Calpouzos, Lucas, University of Idaho
Chatigny, Mark, Naval Biomedical Research Lab, Oakland, California
Dimmick, Robert L., Naval Biomedical Research Lab, Oakland, California
Edmonds, Robert L., Director, University of Washington
Estabrook, George F., University of Michigan
Fink, Jordan N., Wood VA Hospital, Milwaukee, Wisconsin
Flowers, Wayne, University of Michigan
Furniss, Malcolm M., U.S. Forest Service, Moscow, Idaho
Gara, Robert I., University of Washington
Gray, Jane, University of Oregon

Gressitt, J. Linsley, Bernice P. Bishop Museum, Honolulu

Harrington, James B., Environment Canada, Ottawa, Canada

Hasenclever, Herbert F., National Institute of Allergy and Infectious Diseases, Hamilton, Montana

Hibben, Craig R., Kitchawan Research Laboratory, Ossining, New York

Holzapfel, Eugene, Bernice P. Bishop Museum, Honolulu

Hyre, Russell A., USDA, Pennsylvania State University

Kramer, Charles L., Kansas State University

Leedom, John M., University of Southern California

Lighthart, Bruce, Environmental Protection Agency, Corvallis, Oregon

Loosli, Clayton G., University of Southern California

Machta, Lester, NOAA, Washington, D.C.

Marlatt, William E., Colorado State University

Mason, Conrad J., University of Michigan

McCoy, Randolph E., University of Florida, Fort Lauderdale, Florida

McManus, Michael L., U.S. Forest Service, Hamden, Connecticut

Meredith, Donald S., University of Hawaii

Morton, H., University of Michigan

Nichols, Harvey, University of Colorado

Pack, Donald H., NOAA, Washington, D.C.

Quentin, George H., University of Texas, Odessa

Raynor, Gilbert S., Brookhaven National Laboratory

Safferman, Robert S., University of Cincinnati, Ohio

Salvaggio, John, Louisiana State University

Schlicting, Harold E., Jr., Bio Control Company, Port Sanilac, Michigan

Schmidt, Robert A., University of Florida

Shy, Carl M., National Environmental Research Center, Research Triangle Park, North Carolina

Smith, Paul E., Westinghouse Electric Corporation, North Carolina

Snyder, William C., University of California, Berkeley

Solomon, Allen M., University of Arizona

Solomon, William R., University of Michigan

Spendlove, Clifton J., U.S. Army, Dugway Proving Ground, Utah

Thompson, Jack E., National Environmental Research Center, Research Triangle Park, North Carolina

Vale, Gabor, University of Wyoming

Wallin, Jack R., USDA/ARS, University of Missouri

Webb, Thompson, III, Brown University, Rhode Island

MARINE MAMMALS

DIRECTORATE: Director, G. Carleton Ray; Program Administrator, Suzanne Contos

DURATION OF PROGRAM: 1971-72

OBJECTIVES: To provide basic information needed for rational international management of marine mammals and to encourage increased international collaboration among marine mammalogists.

RESULTS: Arranged International Conference on the Biology of Whales in 1971. Published "Marine Mammals Newsletter." Operated "Marine Mammals Office at the U.S. National Museum.

PARTICIPANTS
Bartholomew, George A., University of California, Los Angeles
Buhler, Donald R., Oregon State University
Burns, John J., Alaska Department of Fish and Game, Fairbanks
Caldwell, David K., Communication Sciences Laboratory, Florida
Caldwell, Melba C., Communication Sciences Laboratory, Florida
Chapman, Douglas G., University of Washington
Contos, Suzanne, M., Smithsonian Institution, Washington, D.C.
Cummings, William C., Naval Undersea R & D Center, San Diego
Dailey, Murray D., California State College, Long Beach
Dohl, Thomas P., The Oceanic Institute, Oahu, Hawaii
Doyle, William T., University of California, Santa Cruz
Dyr, Michael A., University of California, Los Angeles
Evans, William E., Naval Undersea R & D Center, San Diego
Farrell, Keith R., Washington State University, Pullman
Fay, Francis H., Arctic Health Research Center, College, Alaska
Fiscus, Clifford, National Marine Fisheries Service, Seattle
Fish, James F., Naval Undersea R& D Center, San Diego
Gentry, Roger L., University of California, Santa Cruz
Goering, John J., University of Alaska
Goodman, Robert M., The Franklin Institute, Philadelphia
Griggs, Gary B., University of California, Santa Cruz
Hall, John D., University of California, Santa Cruz
Handley, Charles O., Jr., Smithsonian Institution, Washington, D.C. .
Harbo, Samuel J., Jr., University of Alaska, Fairbanks
Houch, Warren J., California State University
Irvine, A. Blair, University of Florida
Kenyon, Karl W., U.S. Fish and Wildlife Service, Seattle
Kim, Ke Chung, Pennsylvania State University
Kokjer, Kenneth J., University of Alaska
Leatherwood, J. Stephen, Naval Undersea R&D Center, San Diego

LeBoeuf, Birney J., University of California, Santa Cruz
Lent, Peter C., University of Alaska
Major, Diana M., Fisheries Research Board of Canada, Quebec
Mate, Bruce R., Oregon State University
McRoy, C. Peter, University of Alaska
Miller, L. Keith, University of Alaska
Morejohn, G. Victor, Moss Landing Marine Laboratory, California
Mueller, George J., University of Alaska, College
Newby, Terrell C., University of Washington
Norris, Kenneth S., University of California, Los Angeles
Odell, Daniel K., University of California, Los Angeles
Patten, Donald R., Los Angeles County Museum of Natural History
Paulik, Gerald J., University of Washington
Pearse, J. S., University of California, Santa Cruz
Perrin, William F., Southwest Fisheries Center, La Jolla, California
Peterle, Tony J., Ohio State University
Pierce, Richard W., University of California, Santa Cruz
Prescott, John H., Marineland of the Pacific, Palos Verdes
Ray, Clayton E., Smithsonian Institution, Washington, D.C.
Ray, G. Carleton, Director, Johns Hopkins University, Maryland
Rice, Dale W., Southwest Fisheries Center, La Jolla, California
Rich, Sigmund T., 814 Teakwood Road, Los Angeles
Schevill, William E., Woods Hole Oceanographic Institution
Schusterman, Ronald J., California State College, Hayward
Stoker, Sam W., University of Alaska, College
Sult, Steven D., The Johns Hopkins University, Maryland
Vania, John S., Alaska Department of Fish and Game, Fairbanks
Walker, William A., Marineland of the Pacific, Palos Verdes
Wartzok, Douglas, The Johns Hopkins University, Maryland
Watkins, William A., Woods Hole Oceanographic Institution
Whittow, G. Causey, University of Hawaii
Witt, James M., Oregon State University

L. Creech

DURATION OF PROGRAM: 1971-72

OBJECTIVES: To conduct national surveys of primitive and wild plant genetic resources held in the United States and to determine the scope, quality, and usefulness of these resources.

RESULTS: 9 papers in open literature (to June 1973)

PARTICIPANTS
Brown, William R., Pioneer Hi-Bred Corn Company, Des Moines, Iowa
Chang, T. T., International Rice Research Institute
Creech, John L., Director, U.S. National Arboretum, Washington, D.C.
Harlan, Jack R., University of Illinois
Johnson, Russel T., Spreckels Sugar Company, San Francisco, California
Konzak, Calvin F., Washington State University
Reitz, Louis, ARS, USDA, Beltsville, Maryland

CONSERVATION OF ECOSYSTEMS

DIRECTORATE: Director, John Buckley, 1968-70; Director, George S. Sprugel, 1970-73; Director, Rezneat M. Carnell, 1973-74

DURATION OF PROGRAM: 1968-75

OBJECTIVES: To define a national system of ecological preserves, including development of a classification system for the ecological areas of the United States, inventorying the types of areas already preserved, and locating suitable examples of unrepresented types for inclusion in the system.

RESULTS: 7 papers in open literature (to June 1973); two computer tapes at the Smithsonian Institution containing descriptions of 2,753 terrestrial, freshwater, and marine areas which should be preserved.

PARTICIPANTS
Ash, Grant, U.S. Department of the Army
Bader, Richard, University of Miami (deceased), Florida
Beeton, Alfred M., University of Wisconsin, Milwaukee
Bridge, David, Smithsonian Institution, Washington, D.C.
Buckley, John M., U.S. Environmental Protection Agency
Buckman, Robert E., U.S. Department of Agriculture
Butler, Robert L., Pennsylvania State University
Campbell, Howard W., U.S. Department of the Interior
Carnell, Rezneat M., Texas A&M University
Cole, Gerald A., Arizona State University
Cooper, Arthur W., State of North Carolina, Raleigh
Danstedt, Rudolph T., 6202 Landon Lane, Bethesda, Maryland
Dorris, Troy C., Oklahoma State University
Erdman, Kimball S., Slippery Rock State College, Pennsylvania
Fosberg, F. Ray, Smithsonian Institution, Washington, D.C.
Frank, Peter W., University of Oregon
Franklin, Jerry F., U.S. Department of Agriculture
Goodwin, Mrs. M. M., U.S. Department of the Navy
Heed, William B., University of Arizona
Hirsch, Alan, U.S. Department of Commerce
Holden, David J., South Dakota State University
Jenkins, Dale. W., Smithsonian Institution, Washington, D.C.
Jenkins, Robert, The Nature Conservancy, Arlington, Virginia
Keilley, Steve, Smithsonian Institution, Washington, D.C.
Kibby, Harold, U.S. Environmental Protection Agency
King, Donald R., U.S. Department of State
Lawson, William, Smithsonian Institution, Washington, D.C.
Lemon, Paul C., American Institute of Biological Sciences

Likens, Eugene E., Cornell University, New York
Linn, Robert M., U.S. Department of the Interior
McHugh, J. L., State University of New York, Stony Brook
Maciolek, John A., U.S. Department of the Interior
Mathias, Mildred, University of California, Los Angeles
Milliger, Larry E., Greiner Corporation, Baltimore, Maryland
Morrow, James E., University of Alaska
Neuhold, John M., Utah State University
O'Connor, Dennis, University of Miami, Florida
Osburn, William S., Jr., U.S. Atomic Energy Commission
Randall, John E., Bernice P. Bishop Museum, Honolulu
Ramsey, John S., Auburn University, Alabama
Ray, G. Carleton, The Johns Hopkins University, Maryland
Reid, George K., Eckerd College
Risser, Paul, University of Oklahoma
Russell, Ms. Patricia, American Institute of Biological Sciences
Siccama, Thomas, Yale University, Connecticut
Sigafoos, Robert S., U.S. Department of the Interior
Smith, Robert, U.S. Department of the Interior
Sprugel, George, Jr., Illinois Natural History Survey
Sullivan, Jack, U.S. Department of Agriculture
Talbot, Lee M., Council on Environmental Quality
Ugolini, Frank, U.S. Department of the Interior
van Hook, Kathy, Regional Planning Association of Western Montana,
    Missoula
Waggoner, Gary S., U.S. Department of the Interior
Walford, Lionel, U.S. Department of the Interior (retired)
Willard, Beatrice, Council on Environmental Quality
Wissing, Thomas E., Miami University, Ohio
Zinn, Donald J., University of Rhode Island

BIOSOCIAL ADAPTATION OF MIGRANT AND URBAN POPULATIONS

DIRECTORATE: Director, Demitri B. Shimkin, 1968-69; Director, Everett S. Lee, 1969-75

DURATION OF PROGRAM: 1968-75

OBJECTIVES: To find the answers to the following questions: How do migrants differ from nonmigrants in the area of origin? From nonmigrants in the area of destination? From migrants in other migration streams? From migrants of earlier times? What is the impact of migration on the structure of the population at origin? On the structure of the population at destination? On the social and psychological milieu at origin and destination? What is the impact of migration on the psychological and physical state of the migrants? On the psychological and physical state of the migrants' children? What changes in social and economic conditions have the greatest effect on the volume of migration? On the selection of prospective migrants? On the types of migrants' origins and destinations? What are the most important factors in the assimilation of migrants at destination? What are the most important obstacles to the assimilation of migration?

RESULTS: 6 publications in open literature (to June 1973); one book, "Biosocial Impact of Migration and Urbanization," for US/IBP Synthesis Series (planned).

PARTICIPANTS
    Alaska (U.S.)
Fischer, V., University of Alaska
    Appalachia (U.S.)
Brown, James, University of Kentucky
    Argentina
Wilkie, Richard B., University of Massachusetts
    Cuban Refugees
Wooten, C., University of Miami, Florida
    Finland
Haro, T., National Board of Health, Helsinki
    Holmes County Study (U.S.)
Lorenzi, Henry, Lexington, Mississippi
Montgomery, Bernice, Lexington, Mississippi
Shimkin, Demitri B., Harvard University, Massachusetts
    Israel
Katz, Eliku, University of Jerusalem
Sichron, Moshe, Jerusalem, Israel

Scotland

Illsley, R., University of Aberdeen

Sweden

Husen, Thorsten, University of Stockholm

Thailand

Leoprapai, Boonleert, Central Statistical Office

Goldstein, Sidney, Brown University, Rhode Island

Piampiti, S., Chulalongkurn University

Yazoo County Study (U.S.)

Price, Daniel O., University of Texas

Yugoslavia-Montenegro

Cannell, Charles, University of Michigan

Kilibarda, Momcilo, Belgrade, Yugoslavia

Sagen, O. K., U.S. Public Health Service

Savicevic, Miomir-Mito, Belgrade, Yugoslavia

NUTRITIONAL ADAPTATION TO THE ENVIRONMENT

DIRECTORATE: Director, O. Lee Kline

DURATION OF PROGRAM: 1969-74

OBJECTIVES: To guide mankind toward successful adaptation to environmental changes that relate to food. The program on nutritional adaptation to the environment provided central coordination for approximately 30 research studies carried out in several parts of the world. The focus was upon studies by U.S. and foreign investigators working together on a series of nutrition programs.

The broad subjects covered were malnutrition and nutritional status of populations in Alaska, the Arab Middle East, Thailand, India, Indonesia, Puerto Rico, and the U.S.; nutritional requirements of vitamin B_6, iodine, vitamin C, fat, and in relation to the combat soldier; nutrient deficiencies such as endemic goiter in Latin America, nutritional anemia in Southeast Asia, and zinc deficiency in the Middle East; food and food technology involving protein production from cellulose wastes, soy-based foods for the Philippines, peanut protein, fermented foods and fish protein concentrate; food composition of foods of East Asia, aflatoxins as a food toxicant and nutritional value of plant foods.

The program led to important development of nutrition research capability in Asian and Latin American countries, and provided a demonstration of specific effects of malnutrition seen in some countries but not others. As a result of the IBP emphasis on an interdisciplinary approach to the study of complex ecosystems there has been a broader recognition among nutrition scientists that malnutrition is a socioeconomic as well as a nutritional phenomenon.

RESULTS: 115 papers in open literature.

PARTICIPANTS
Tropical and Subtropical Ecology
    Development of food from cellulosic wastes
Callihan, C. D., Louisiana State University
Daly, W. H., Louisiana State University
Dunlap, C. E., Louisiana State University
Han, Y. W., Louisiana State University
McKnight, William F., Louisiana State University
Srinivasan, V. R., Louisiana State University
    Development of soy-based foods of high nutritive value
Bourne, M. C., University of the Philippines
Hackler, L. Ross, Cornell University, New York
Moyer, J. C., Cornell University, New York

Robinson, W. B., Cornell University, New York
Steinkraus, Keith H., Cornell University, New York
Van Buren, J. P., Cornell University, New York
   Effect of undernutrition on nucleic acid and
   protein content of developing brain
Fish, Irving, Cornell Medical College, New York
Meneghello, Julio, Hospital Roberto del Rio, Santiago, Chile
Rosso, Pedro, Hospital Roberto del Rio, Santiago, Chile
Winick, Myron, Cornell Medical College, New York
   Effects of malnutrition in early life
Barnes, Richard H., Cornell University, New York
   Endemic goiter in Latin America
Dodge, Philip, Washington University, St. Louis
Dunn, John T., University of Virginia
Estrella, Eduardo, Escuela Politecnica Nacional, Ecuador
Fierro, Rodrigo, Escuela Politecnica Nacional, Ecuador
Pretell, Eduardo, Instituto de Investigaciones de la Altura, Lima, Peru
Querido, Andries, Dukzigt Hospital, Rotterdam, Holland
Ramirez, Ignacio, Escuela Politecnica Nacional, Ecuador
Stanbury, John B., Massachusetts Institute of Technology
   Evaluation of fish protein concentrate
Chichester, Chester O., University of Rhode Island
Monckenberg, Fernando, University of Chile
   Food composition table for use in East Asia
Butrum, R. R., National Institutes of Health
Wu Leung, W. T., U.S. Department of Health, Education, and Welfare
   Hair root morphology and protein synthesis in disease
Bradfield, Robert B., University of California, Berkeley
Jelliffe, D. B., University of the West Indies
   Improved methods of extracting and utilizing plant proteins
Hackler, L. Ross, Cornell University, New York
Steinkraus, Keith H., Cornell University, New York
van Veen, A. G., Cornell University, New York
   Increased protein supplies through peanut improvement
Bailey, W. K., U.S. Department of Agriculture
   Influence of specific nutritional deficiency
   on the genetic expression of anemia
Bertles, John F., St. Luke's Hospital Center, New York
Milner, Paul F. A., University of the West Indies
   Malnutrition in the Middle East
Darby, William J., Vanderbilt University, Tennessee
Patwardhan, V. N., Vanderbilt University, Tennessee

Multiple etiology of nutritional anemias of Thailand

Allen, Donald, St. Louis University, Missouri

Amatayakul, Kosin, Chiang Mai Medical College, Thailand

Bailey, Gordon B., Bangkok, Thailand

Chinboot, Boonlert, Chiang Mai Medical College, Thailand

Damrongsak, Damri, Chiang Mai Medical College, Thailand

Horwitt, Max N., St. Louis University, Missouri

Kulapongs, Panja, Chiang Mai Medical College, Thailand

Leitzmann, C., Chiang Mai Medical College, Thailand

Morehead, David, St. Louis University, Missouri

Olson, Robert E., St. Louis University, Missouri

Simmersbach, Franz, St. Louis University, Missouri

Srisukri, Avudh, Chiang Mai Medical College, Thailand

Suskind, Robert, Chiang Mai Medical College, Thailand

Suttajit, Maitree, Chiang Mai Medical College, Thailand

Tan, Charles H., St. Louis University, Missouri

Thanangkul, Ousa, Chiang Mai Medical College, Thailand

Vithayasai, Vicharn, Chiang Mai Medical College, Thailand

Whitaker, Jo Ann, St. Louis University, Missouri

Mycotoxin contamination of food and foodstuffs in Southeast Asia

Bhamarapravati, Natth, University of Medical Sciences, Bangkok, Thailand

Christensen, C. M., University of Minnesota

Shank, R. C., Massachusetts Institute of Technology

Wogan, G. N., Massachusetts Institute of Technology

Nutrition and infections in India

Kielmann, Arnfried, Punjab, India

Ramalingaswami, V., Indian Council of Medical Research, New Delhi

Reinke, William A., The Johns Hopkins University, Maryland

Taylor, Carl E., The Johns Hopkins University, Maryland

Uberoi, I. S., Punjab, India

Wahi, P. N., Indian Council of Medical Research, New Delhi

Nutrition research in Indonesia and Thailand

Gyorgy, Paul, Philadelphia General Hospital

Prawiranegara, Dradjat D., University of Indonesia

Valyasevi, Aree, Ramathibodi Hospital, Bangkok, Thailand

Van Reen, R., University of Honolulu

Nutritional anemia in Southeast Asia

Herbert, Victor, Mt. Sinai School of Medicine, New York

Nutritional studies in Puerto Rico

Asenjo, Conrado F., University of Puerto Rico

Burgos, Jose C., University of Puerto Rico

Fernandez, Nelson A., University of Puerto Rico

Relationship of diet to the performance of the combat soldier
Consolazio, C. Frank, Fitzsimons General Hospital, Denver, Colorado
Daws, T. A., Fitzsimons General Hospital, Denver, Colorado
Johnson, H. L., Fitzsimons General Hospital, Denver, Colorado
Krzywicki, H. J., Fitzsimons General Hospital, Denver, Colorado
Role of zinc deficiency in growth retardation
Halstead, James A., Hudson, New York
Reinhold, John G., Phalavi University, Iran
Smith, James C., Veterans Administration
Diet and fat metabolism
Adams, Mildred, U.S. Department of Agriculture
Benton, Duane A., Ross Labs, Ohio
Experimental scurvy in man
Hodges, Robert E., University of California, Davis
*Human iodine requirements and substances
in foods that affect thyroid function*
Adams, Mildred, U.S. Department of Agriculture
Benton, Duane A., Ross Labs, Ohio
Leverton, Ruth M., U.S. Department of Agriculture
Nutritional status of preschool children in the United States
Owen, George M., Ohio State University
Nutritional value of wheat and wheat products
Gortner, W. A., U.S. Department of Agriculture
Toepfer, E. W., U.S. Department of Agriculture
Nutritive factors in the immunological process
Axelrod, A. E., University of Pittsburgh, Pennsylvania
Protein value of plant foods
Adams, Mildred, U.S. Department of Agriculture
Kelsay, June, U.S. Department of Agriculture
Leverton, Ruth M., U.S. Department of Agriculture
Toepfer, E. W., U.S. Department of Agriculture
*Relationship of dietary deficiency and impairment
of mental and somatic development*
Anderson, D. C., University of Notre Dame, Indiana
Pleasants, J. R., University of Notre Dame, Indiana
Reddy, B. S., University of Notre Dame, Indiana
Wostmann, Bernard S., University of Notre Dame, Indiana
Vitamin B_6 metabolism in man
Linkswiler, Hellen, University of Wisconsin

Arctic Ecology
Nutrition survey among Eskimos of Wainwright, Alaska
Draper, Harold H., University of Illinois
Goad, W., Fitzsimons General Hospital, Denver, Colorado

Herman, Y. F., Fitzsimons General Hospital, Denver, Colorado
Jamison, Paul L., Indiana University
Milan, Frederick A., University of Alaska
Sauberlich, Howard E., Fitzsimons General Hospital, Denver, Colorado
  Nutrition of Eskimos in northern Alaska
Bell, R. R., University of Illinois
Bergan, J. G., University of Illinois
Colbert, M. J., University of Illinois
Draper, Harold H., University of Illinois
Feldman, S. A., Northwestern University, Illinois
Heller, Christine A., U.S. Public Health Service (retired)
Hursh, Laurence M., University of Illinois
Mann, George V., Vanderbilt University, Tennessee
Wo, Catherine W., University of Illinois

Appendix C

Acronyms Used

AAAS	American Association for the Advancement of Science
AEC	Atomic Energy Commission
AFOSR	Air Force Office of Scientific Research
AIBS	American Institute of Biological Sciences
AID	Agency for International Development
ALAIH	Asociación Latinoamericana de Ictólogos y Herpetólogos
AOE	Analysis of Ecosystems
ASA	American Society of Agronomy
ASOVAC	Asociación Venezolana para el Avance de la Ciencia
BOB	Bureau of the Budget
C/B	Cost/Benefit
CE	Conservation of Ecosystems (in US/IBP)
CEQ	Council on Environmental Quality
CES	Committee on Ecosystem Studies
CIA	Central Intelligence Agency
CIBP	Committee for the International Biological Program
CNP	Conselho Nacional de Pesquisas
COPAWE	Commission for Predictive Analysis of World Ecosystems
CORP	Coordinated Research Program
CT	Conservation, Terrestrial (section of IBP)
EM	Environmental Management
EP	Environmental Physiology (in US/IBP)
ERTS	Earth Resources Technical Satellite
ESA	Ecological Society of America
FAO	Food and Agriculture Organization
FASEB	Federation of American Societies for Experimental Biology
FCST	Federal Council on Science and Technology
FY	Fiscal Year
GEMS	Global Environmental Monitoring System
GMS	Global Monitoring System
GNEM	Global Network for Environmental Monitoring
HA	Human Adaptability (section of IBP)
HCR	House Concurrent Resolution
HEW	Department of Health, Education, and Welfare
HJR	House Joint Resolution
IAIE	Interamerican Institute of Ecology

259

IBP	International Biological Program
ICC	Interagency Coordinating Committee
ICE	International Center for the Environment
ICSU	International Council of Scientific Unions
IEPC	International Environmental Programs Committee
IGY	International Geophysical Year
INCORA	Instituto Colombiano de la Reforma Agraria
INDERENA	Instituto de Desarrollo de los Recursos Naturales Renovables
INPA	Instituto Nacional de Pesquisas de Amazonia
INTA	Instituto Nacional de Tecnologia Agropecuaria
IRP	Integrated Research Program
IUBS	International Union of Biological Sciences
IUCN	International Union for Conservation of Nature and Natural Resources
IUGG	International Union of Geodesy and Geophysics
IUPAC	International Union of Pure and Applied Chemistry
JIBP	Japanese Committee for the International Biological Program
LDC	Lesser Developed Country
MAB	Man and the Biosphere (UNESCO program)
MIWEE	Man's Interaction with Earth's Ecosystems
NAS	National Academy of Sciences
NASA	National Aviation and Space Administration
NEL	National Environmental Laboratory
NIH	National Institutes of Health
NRC	National Research Council
OAS	Organization of American States
OECD	Organization for Economic Cooperation and Development
OST	Office of Science and Technology
OTS	Organization for Tropical Studies
PAC	Public Affairs Council
PAWE	Predictive Analysis of World Ecosystems
PF	Productivity Freshwater (section of IBP)
PM	Productivity Marine (section of IBP)
PROCOM	Program Coordinating Committee
PSAC	President's Science Advisory Committee
PT	Productivity Terrestrial (section of IBP)
R & D	Research and Development
SCAR	Scientific Committee on Antarctic Research
SCIBP	Special Committee for the International Biological Program
SCOPE	Scientific Committee on Problems of the Environment
SCOR	Scientific Committee on Oceanic Research

SJR	Senate Joint Resolution
SPVEA	Superintendency of Economic Valorization Plan of Amazonia
SR	Senate Resolution
TIE	The Institute of Ecology
UN	United Nations
UNEP	United Nations Environmental Program
UNESCO	United Nations Educational, Scientific, and Cultural Organization
USAF	United States Air Force
USEC/IBP	U.S. Executive Committee, International Biological Program
USNC/IBP	U.S. National Committee, International Biological Program
WHO	World Health Organization